Caribbean Primary Agriculture

Book 4

New Edition

R. Ramharacksingh
DipAgric; AdvDip Tech'l Teacher Trng;
BSc (Hons) Agric; MA Agric'l Educ'n

Illustrated by G. J. Galsworthy and Donald Mullis

Great Clarendon Street, Oxford, OX2 6DP, United Kingdom

Oxford University Press is a department of the University of Oxford. It furthers the University's objective of excellence in research, scholarship, and education by publishing worldwide. Oxford is a registered trade mark of Oxford University Press in the UK and in certain other countries

Text © R. Ramharacksingh 1982, 1999

The moral rights of the authors have been asserted

First published in 1982
Second edition published in 1999
Reprinted in 2000 by Continuum
Reprinted in 2002 by Nelson Thornes Ltd
This edition published by Oxford University Press in 2014

All rights reserved. No part of this publication may be reproduced, stored in a retrieval system, or transmitted, in any form or by any means, without the prior permission in writing of Oxford University Press, or as expressly permitted by law, by licence or under terms agreed with the appropriate reprographics rights organization. Enquiries concerning reproduction outside the scope of the above should be sent to the Rights Department, Oxford University Press, at the address above.

You must not circulate this work in any other form and you must impose this same condition on any acquirer

British Library Cataloguing in Publication Data
Data available

978-0-7487-7219-3

10 9 8

Printed in India by Multivista Global Pvt. Ltd.

Although we have made every effort to trace and contact all copyright holders before publication this has not been possible in all cases. If notified, the publisher will rectify any errors or omissions at the earliest opportunity.

Links to third party websites are provided by Oxford in good faith and for information only. Oxford disclaims any responsibility for the materials contained in any third party website referenced in this work.

Books in this series
Level One (5–7 years)
Workbook for Infants

Level Two (7–9 years)
Textbook 1 Workbook 1

Level Three (9–11 years)
Textbook 2 Workbook 2

Level Four (11–15 years)
Textbook 3 Workbook 3
Textbook 4 Workbook 4

The Author

The author of the Caribbean Primary Agriculture series, Mr Ron Ramharacksingh, is professionally qualified and a very experienced educator, trainer and agriculturist.

He has served as a trained primary school teacher, lecturer and teacher trainer: teachers' training colleges; Assistant Director: agriculture teacher education and training; FAO Consultant: curriculum development; and Curriculum Officer: agricultural education.

He is currently serving as Supervisor, Technical Teacher Training, Ministry of Education, Trinidad and Tobago, West Indies.

The manufacturer's authorised representative in the EU for product safety is Oxford University Press España S.A. of El Parque Empresarial San Fernando de Henares, Avenida de Castilla, 2 – 28830 Madrid (www.oup.es/en or product.safety@oup.com). OUP España S.A. also acts as importer into Spain of products made by the manufacturer.

Preface

This series has been specially written to meet the needs and growing interest in agricultural science in Caribbean primary schools.

Primary Agriculture is an integral part of the school curriculum in most countries of the Caribbean, a region where agriculture and tourism are major sectors of the economy.

This revised edition of the Caribbean Primary Agriculture series provides technological update of the various units and topics, which are comprehensively sequenced and graded to suit the various age groups and class levels of the primary school. The Scientific Discovery Approach is used so that students can find solutions to problems both by their own investigations and with guidance from the teacher.

Primary Agriculture focuses on awareness, orientation and exploration. The main general objectives of the programme should enable students to:

(i) understand that agriculture is an art or a skill, a science, a business and a vocation;
(ii) develop an awareness of the importance of agriculture to the national economy and the environment;
(iii) acquire the prerequisites for pursuing agriculture at a higher level.

For more effective teaching and learning, teachers need to use a variety of methods, techniques and strategies and practise integration and correlation of agriculture with every subject area of the primary school curriculum. The multi-faceted nature of agriculture provides several diverse interests and topics which can be used as centres of interest and pursued by students as group or individual projects.

The activities in the Workbooks are geared primarily to reinforce theoretical knowledge and problem-solving skills. They must be supplemented with relevant demonstrations and participation by students in practical agricultural activities both at school and at home.

In planning lessons, teachers should adapt the textbook material and select appropriate examples suited to their environment.

Textbooks and Workbooks 3 and 4 are useful resource materials for students pursuing agriculture at Post Primary Centres, secondary schools (Forms 1 to 3), Youth Development and Apprenticeship Centres (former Youth Camps), farm schools and teachers' training colleges.

R. Ramharacksingh

Contents

Unit 1 **Preparations before planting**
1. Land preparation practices 7
   - (a) Land clearing 7
   - (b) Tillage 9
   - (c) Manuring 11
   - (d) Liming 12
   - (e) Drainage 13
   - (f) Levelling 14
2. Preparation of pots, tins, tyres and troughs 14
3. Preparation of a potting soil 16
4. Nutriculture and hydroponics 19
5. Grow-boxes 20
6. Seedboxes, seedbeds and speedling trays 20
   - (a) Seedboxes 20
   - (b) Seedbeds 22
   - (c) Speedling trays 23

Unit 2 **Selection and management of planting materials**
1. Plant propagation 24
   - (a) Natural methods of plant propagation 24
   - (b) Artificial methods of plant propagation 25
2. Planting materials 28
   - (a) Types of planting materials 28
   - (b) Selection of planting materials 28
   - (c) Nursery techniques or practices 29
   - (d) Management or care of young plants 30
   - (e) Procedures for planting out 32

Unit 3 **Field management of crops**
1. Cultural practices 35
2. Drainage and irrigation 38
   - (a) Drainage 38
   - (b) Irrigation 39
3. Pests and diseases 41
   - (a) Pests 41
   - (b) Diseases 42
   - (c) Methods of controlling pests and diseases 45
4. Manuring and fertilising 48
   - (a) Nutrients required by plants 48
   - (b) Supplying nutrients to plants 48
   - (c) Manures or nutrient suppliers 49
   - (d) The nitrogen cycle 52
   - (e) Storage of manures 53
   - (f) Application of manures 54
5. Weed control 55

Unit 4 **Crop production**
1 Factors affecting crop production 58
2 Steps for growing a crop 58
3 Major groups of crops 59
4 Cultivation of crops 61
  Crop Guide 66

Unit 5 **Livestock rearing**
1 Factors affecting livestock rearing 70
2 Breeds and breeding 70
   (a) Breeds 70
   (b) Breeding 71
   (c) The importance of breeding 72
   (d) Breeds of animals 73
3 Care of young farm animals 73
   (a) Chicks 74
   (b) Rabbits 76
   (c) Goats 77
4 Management of growing and adult livestock 78
   (a) Management of broilers 78
   (b) Management of layers 80
   (c) Management of rabbits 81
   (d) Management of goats 83
5 Bee-keeping 84
   (a) The hive 84
   (b) The economic importance of bee-keeping 85

Unit 6 **The business of agriculture**
1 Agriculture as a business 86
   (a) Factors and main objectives 86
   (b) Importance of planning 86
   (c) Simple financial records 88
2 Marketing of produce 89
3 Industry and agriculture 91
   (a) How industry helps agriculture 91
   (b) How agriculture helps industry 91

Unit 7 **Agencies which help farmers**
Government and non-government agencies 93

Unit 1 **Preparations before planting**

1 Land preparation practices

(a) Land clearing
Look at the pictures below:

Figure 1   *Figure 2*

Figure 1 shows a school garden after the long July–August vacation. Figure 2 shows a large corn field after the crop has been harvested. Observe the tall grass, bushes, remains of crop plants and stakes in the two pictures.

What is the first operation that must be done before a new crop is planted? Before growing a new crop, the land must be cleared of all grass, bushes, stakes and crop residues such as the corn stalks.

Can you say what method of land clearing should be used in each case? Study the methods of land clearing given below before you decide. Give reasons for your decisions.

Manual method of land clearing   *Mechanical method of land clearing*

Operation 1
Cutlassing, slashing, brushcutting
Chopping, hoeing (weeding),
pulling out stakes

Operation 1
Brushcutting

Operation 2
Heaping, stacking, windrowing, raking

Operation 2
Heaping, stacking, windrowing, raking

If the grass, bush and crop residues are not too plentiful and will not interfere with ploughing, it is advisable to spread the cut material over the land and plough it into the soil.
Can you explain the advantages or benefits of doing this? The cut material adds organic matter to the soil which improves soil structure and fertility.

Now look at the picture below. It is a virgin forest.

Observe the tall trees with thick trunks, the twining lianes and the dense canopy.

Can you explain how land clearing of a virgin forest is done? Find out by studying the steps given below:

Manual method
Operation 1
Cutting, felling, logging, lopping of trees

Mechanical method
Operation 1
Bulldozing trees

Operation 2
Removal of valuable logs

Operation 2
Removal of valuable logs

Operation 3
Heaping, stacking, windrowing, raking

Operation 3
Heaping, stacking, windrowing, raking

Can you explain why it is not a good idea to use burning as a method of land clearing?
It is because burning destroys the valuable organic matter on the soil surface and exposes the bare soil to erosion. Useful soil organisms (macro and micro) are also destroyed.

It is better, therefore, to allow the cut plant materials, stacked in heaps and windrows, to rot. What are some of the advantages of this practice?

The plant material on the soil surface helps to prevent soil erosion. As the organic matter decomposes or rots, nutrients are slowly released to the soil. In this way the land remains fertile over a longer period.

Now turn to your Workbook and do **Activity 1**.

(b) Tillage
(i) Ploughing or primary tillage
Ploughing is the breaking up of the soil using manual or mechanical tools and equipment.

Manual tools for ploughing

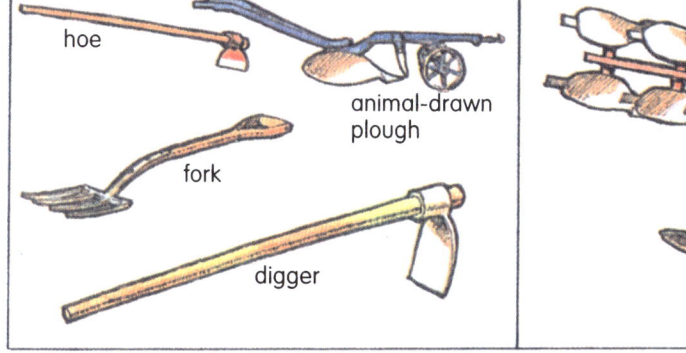

Mechanical equipment for ploughing

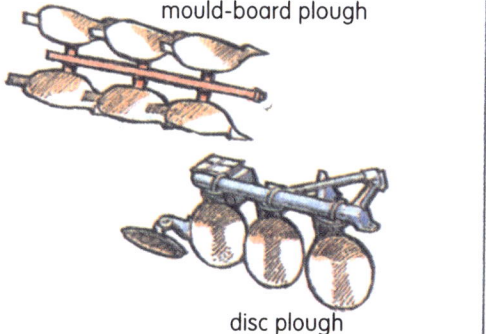

Carefully observe the changes that take place to the soil surface as land is being ploughed. From your observations, can you list some advantages or benefits of ploughing? The diagrams below will help you.

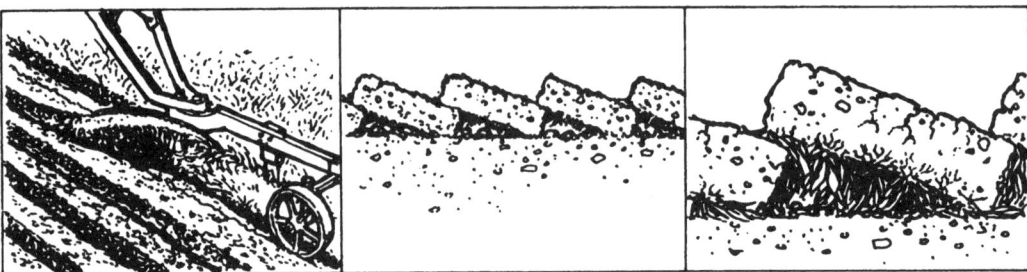

1 Ploughing loosens or breaks up the soil surface.
2 Ploughing enables air and water to enter the soil more freely.
3 Ploughing helps to bury, mix or incorporate organic matter such as grass, bush, weeds and crop residues in the soil.

(ii) Refining or secondary tillage
Refining is the breaking up of the large lumps of soil after it has been ploughed into smaller clods or aggregates.

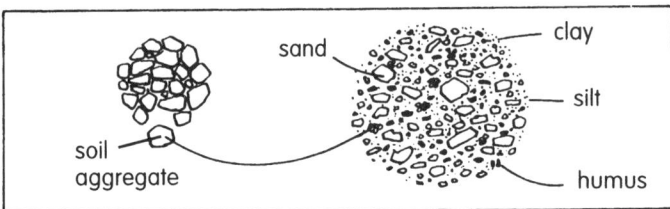

Soil aggregates: each aggregate (piece of soil) consists of a mixture of sand, silt, clay and humus particles.

Look at the diagrams below and find out the names of tools and equipment which are used for refining the soil.

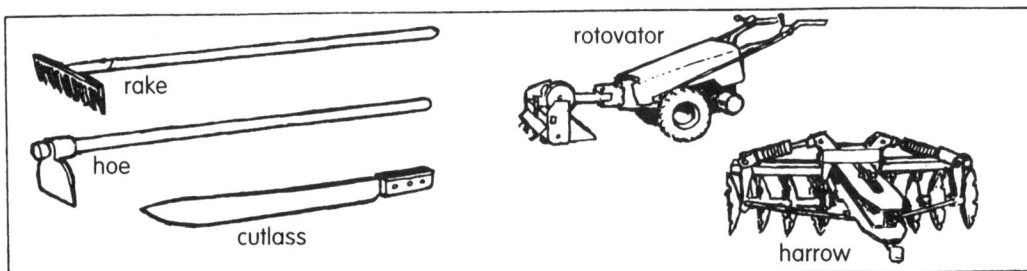

Can you explain why it is necessary to refine the soil after ploughing?
— to obtain a suitable *tilth* (that is, the breaking up and proper mixing of soil particles to form *fine* or *medium-sized* clods or aggregates);
— to obtain a proper seed-bed for growing crops;
— to enable the roots of crop plants to penetrate the soil more freely;
— to cut up, mix or incorporate organic matter, such as crop residues, in the soil.

Now do **Activity 2** in your Workbook.

(c) Manuring

At land preparation time, manure (organic and inorganic) may be spread or broadcast over the ploughed land and rotavated into the soil.

Give some examples of organic and inorganic manures. The chart below will help you.

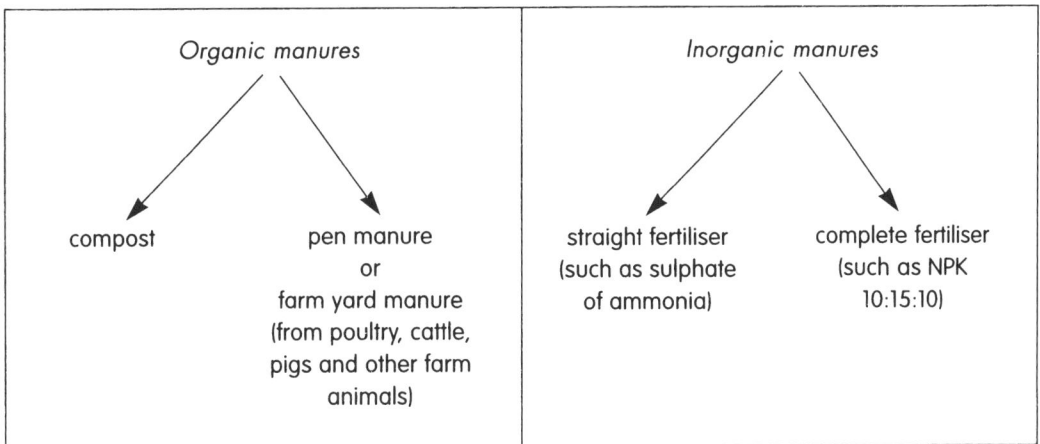

Can you name some tools and equipment which are used for applying manure to ploughed land? The diagrams below will help you.

What are some benefits of incorporating manure in the soil at land preparation time?
— Manure supplies nutrients to crop plants.
— With nutrients readily available in the soil, crop plants gain a head-start and grow more vigorously than crops to which manures were added after planting.
— Organic manure holds soil moisture for crop plants.
— Manure enriches the soil.
— Manure improves soil structure.
— Manure helps to keep or maintain the soil in a fertile state.

(d) Liming

It may be necessary to broadcast and incorporate agricultural lime into the soil at land preparation time. Can you say why?

The heavy rainfall which occurs in the Caribbean Islands, and tropical countries generally, causes most of the soluble nutrients such as *calcium* and *magnesium* to be washed out or leached from the soil. The soil then becomes *too acid* for crop plants to grow healthily and yield properly.

How can you tell if a particular soil is too acid?
A soil test is carried out to determine whether the soil is too acid or not.
In most Caribbean countries soil testing is done as a free service to farmers by the Ministry of Agriculture. A soil test determines the pH (hydrogen potential) of a particular soil.

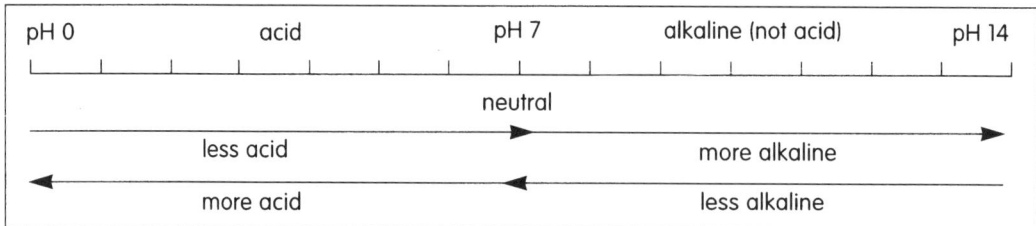

The diagram above shows a pH scale. Soils which have a pH of below 7 are acid whereas those having a pH of above 7 are alkaline (or not acid).
When a soil test is taken and the soil is too acid, the farmer is advised to add a certain quantity of agricultural lime to his land, for example, five hundred kilograms per hectare (500 kg/ha).

When the soil is too acid, what effect does it have on crop plants?
When the soil is too acid, for example pH 4.0, crop plants are not able to absorb or use up the various nutrients from the soil even though these nutrients are present in the soil.
However, when lime is used as recommended, it reduces the acidity of the soil, for example, from pH 4.0 to pH 6.5.

What are some of the benefits of liming?
Generally, liming is done to bring about a pH range of 6.0 to 6.5.
At this pH range, crop plants are able to absorb and utilise the various nutrients present in the soil and grow healthily and yield abundantly.
Liming, therefore, helps to maintain the soil in a fertile state.

Can you name some liming materials?
Liming materials are generally referred to as '*soil-sweeteners*'.
Some liming materials are as follows:
— *calcium oxide* (also known as quicklime or burned lime)
— *calcium hydroxide* (also known as slaked lime)
— *calcium carbonate* (also known as calcite)
— *calcium magnesium carbonate* (also known as dolomite).

Liming materials which are commonly used are calcite and dolomite, and these are known as ground limestone. As their names suggest, calcite (*calcium carbonate*) supplies only calcium to the soil whereas dolomite (*calcium magnesium carbonate*) supplies both calcium and magnesium.
Of the two types of ground limestone, calcite and dolomite, which do you think is better? Why?

Now turn to your Workbook and do **Activity 3**.

(e) Drainage

Drainage is the removal of excess water from the soil surface as well as from within the soil or sub-surface.

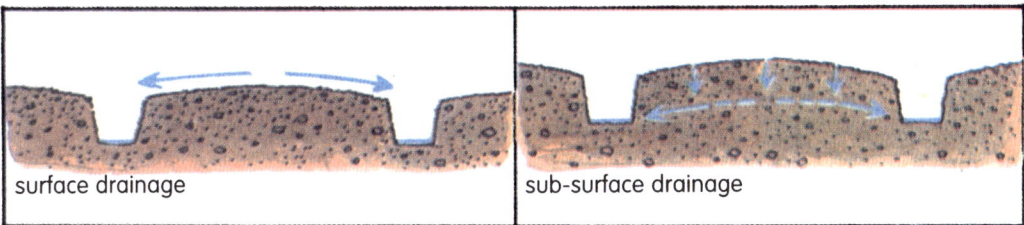

Name some tools and equipment which are used for drainage.

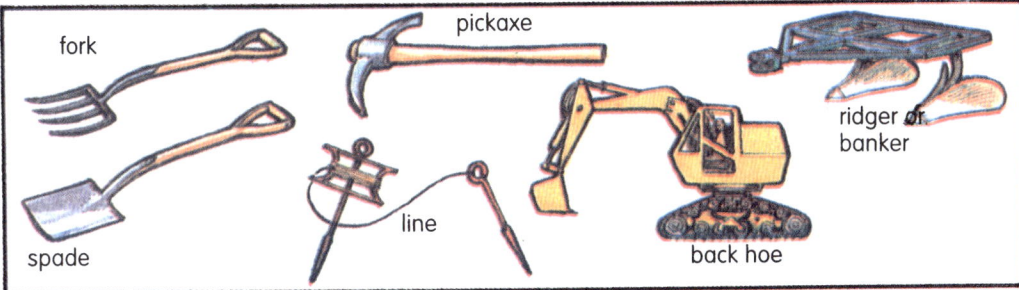

Special types of bed may be formed at land preparation time to make drainage easier.

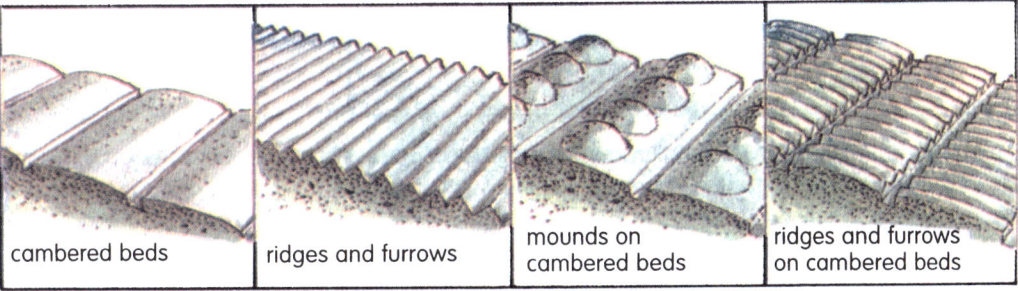

Can you name some types of drains? The diagrams below will help you.

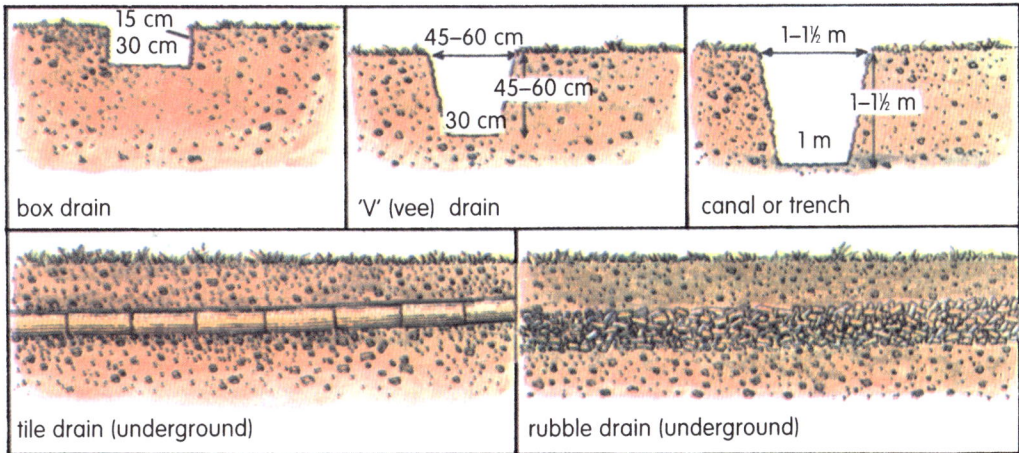

What are some of the benefits of drainage?
— drainage enables the quick removal of water from the surface and sub-surface of the soil;
— drainage enables air to enter the soil (soil aeration). You have learnt previously that air is needed by the roots at plants as well as by soil organisms;
— drainage removes toxic or poisonous substances from the soil.

You will learn more about drainage in a later lesson.

(f) Levelling
The final step of land preparation is to level the garden beds.
Here are some tools and equipment which are used for levelling:

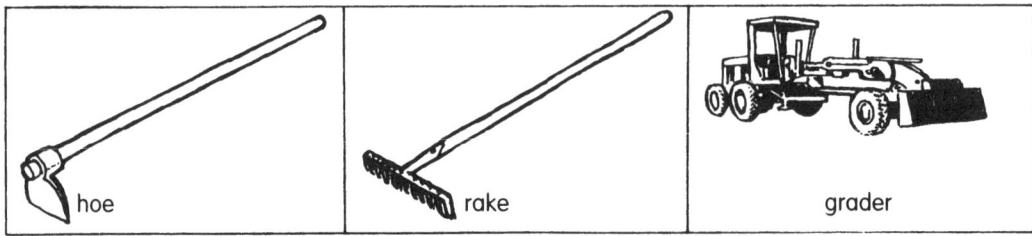

When drains have been dug heaps of soil are piled on the sides of the garden beds.

Apart from giving the garden an untidy appearance, the heaps of soil occupy valuable space where crops should be planted. So these heaps of soil must be spread or levelled off on the garden bed. Whilst levelling, it may be necessary to fill holes or depressions or to cut down mounds and ridges left on the surface after rotavating.

Now do **Activity 4** in your Workbook.

2 Preparation of pots, tins, tyres and troughs
Pots, tins, tyres and troughs are very useful and are specially recommended for homes and schools where suitable land is not available for making garden beds.

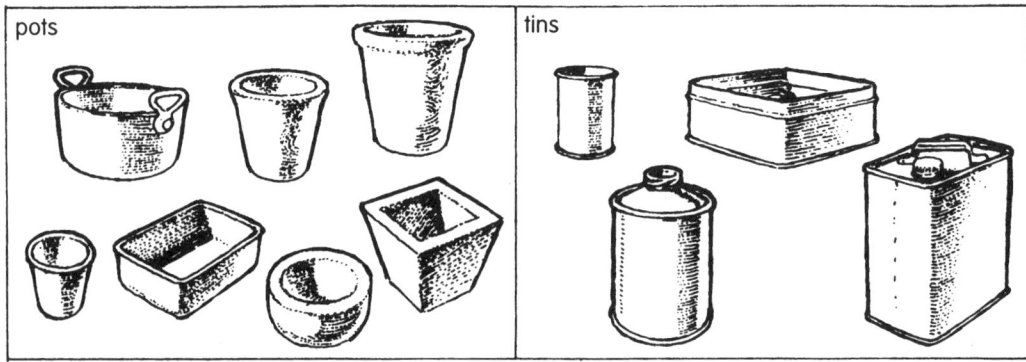

They can readily be obtained or constructed using scrap, discarded and local materials. Some materials which can be used are discarded cooking pots; plastic, clay and concrete pots; large milk tins; biscuit tins; cooking oil tins; barrels; box-wood; bamboo; round-wood; lumber; large PVC drain pipes, bricks and used tyres.

Look at the diagrams below to find out the names of some tools and equipment needed for constructing and preparing containers for growing crops.

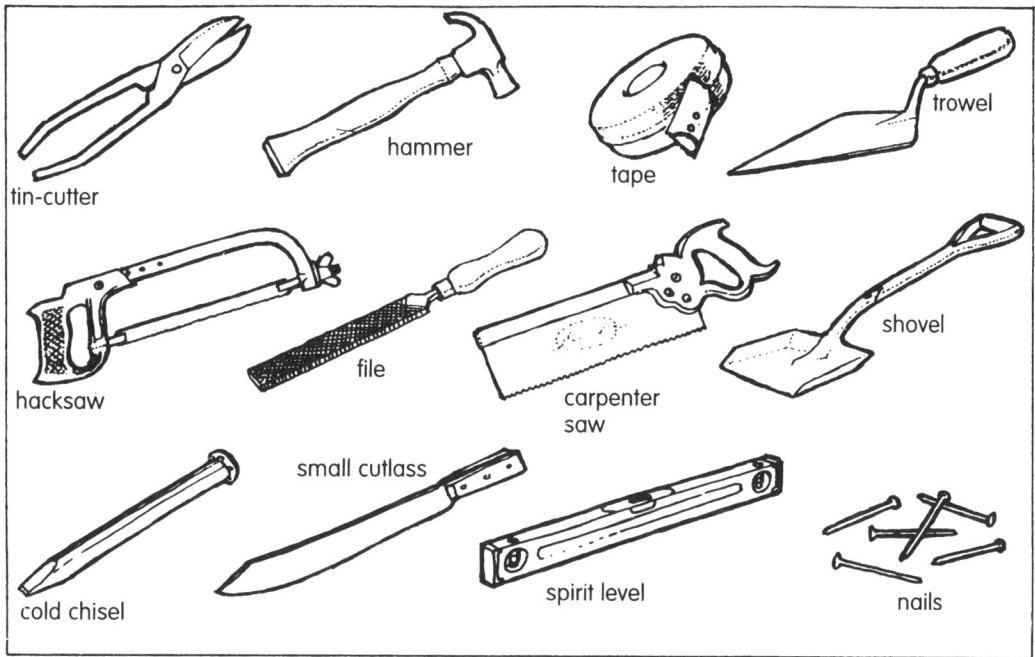

Here are some points to remember when constructing and preparing containers for cultivating crops. Observe and assist your teacher as he demonstrates the steps:

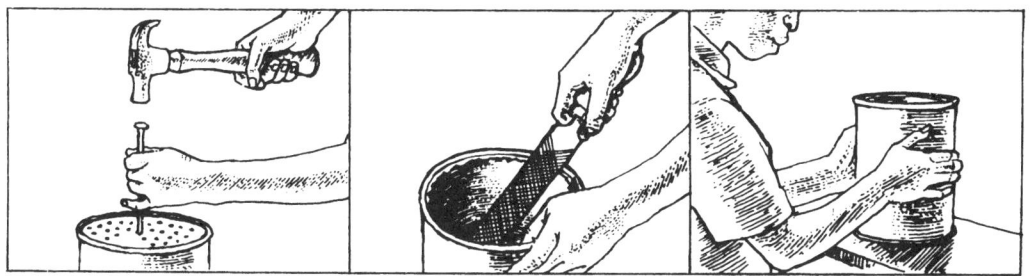

1. Make holes or slits at the base of each container for drainage.
2. Hammer (pound in) or file off sharp edges on tins. Why?
3. Set up or place the containers where they can receive sunlight. Do you know why?
4. Place a layer of straw, dry grass, or coconut leaves on the base inside each container before filling with potting soil. Why?
5. Apply paint to the containers to preserve and make them attractive.

Now turn to your Workbook and do **Activity 5**.

3 Preparation of a potting soil

In the previous lesson you learnt how to construct and prepare containers for the cultivation of crops. Can you state the materials you will need to fill or prepare each container before you actually plant seeds or seedlings? Find out by studying the diagrams below:

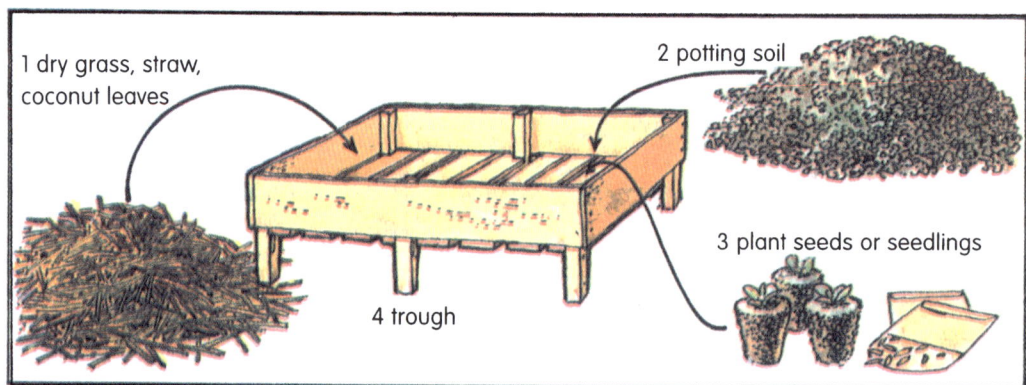

What are the features of a good potting soil?
The list below will help you to find the answers.

Features of a good potting soil
1. has a proper soil aggregation (soil structure)
2. contains essential nutrients for plants
3. provides root room for the free growth of roots
4. is well aerated
5. enables free drainage
6. holds adequate soil moisture
7. is free from harmful pests such as insects, fungi, nematodes
8. is free from recognisable weeds
9. has a proper soil pH (6.0 to 6.5)

Now, name materials which can be used for preparing a potting soil.
What are the main functions of each material?

Materials	Functions in the potting soil mixture
Top soil or clay	1 to supply nutrients 2 to hold soil moisture
Organic manure e.g. pen manure compost manure	1 to supply nutrients 2 to hold soil moisture 3 to produce good soil aggregates (soil structure)
Sharp sand or river sand	1 to enable free drainage 2 to provide soil aeration 3 to provide root room for the free growth of roots
Coconut fibre-bast or rotted bagasse or rotted coffee hull	1 to provide good soil aggregation or soil structure 2 to provide soil aeration 3 to enable free drainage 4 to hold soil moisture 5 to provide root room for the free growth of roots
Inorganic manure e.g. NPK fertiliser (10:15:10)	1 to supply essential plant nutrients such as nitrogen (N), phosphorus (P) and potassium (K)
Ground limestone e.g. dolomite	1 to supply calcium and magnesium to the soil 2 to obtain a proper soil pH (6.0 to 6.5)
Soil insecticide e.g. Diazinon	1 to destroy harmful soil insects and creatures
Fungicide e.g. Banrot, Captan	1 to destroy harmful soil fungi, such as damping-off fungus
Soil sterilant e.g. Formaldehyde, Chloropicrin	1 to destroy soil pests such as insects, fungi, nematodes 2 to destroy weed seeds, weed cuttings

Having obtained the necessary materials, what quantity of each material will you use for preparing a good potting soil?

Several recipes or mixtures can be used. However, with the aid and guidance of your teacher follow the *recipe* and *procedure* as outlined below. They will enable you to produce a good potting soil.

Recipe for a good potting soil
Materials
(a) 3 parts top soil or clay
(b) 2 parts pen manure
(c) 1 part sharp sand or river sand
(d) 2 parts coconut fibre-bast or rotted bagasse or rotted coffee hull
(e) 500 gm NPK (10:15:10) fertiliser per cubic metre (m^3) of potting soil
(f) 500 gm ground limestone (dolomite) per cubic metre (m^3) of potting soil
(g) 30 ml soil insecticide (Diazinon) per 4.5 l of water
(h) 10 gms fungicide (Banrot) per 4.5 l of water
(i) 60 ml soil sterilant (Formaldehyde) per 4.5 l of water

Procedure

1. Mix items (a) to (f) thoroughly using a shovel.
2. Mix items (g) and (h) together in 4.5 l of water in a spray can.
3. Spread out soil mixture thinly and spray with insecticide/fungicide mixture.
4. Spray the soil mixture with the soil sterilant (Formaldehyde mixture).
5. Heap soil, cover over with a large cellophane sheet and leave it covered for 2–3 days. This enables the chemicals to be more effective in destroying soil pests.
6. After 2–3 days, remove the cellophane cover, spread out the soil mixture thinly and allow it to aerate for 1–2 days. This gets rid of poisonous fumes or gases.
7. After 1–2 days, sift the potting soil using a sieve made with a one centimetre square (1 cm^2) wire mesh. Find out why sifting is done.
8. Store the sifted potting soil in a sheltered area protected from rain. Why?
9. Use the potting soil to fill containers – seed boxes, pots, tins, tyres and troughs.

Now do **Activity 6** in your Workbook.

4 Nutriculture and hydroponics

The diagram below shows tomato plants growing in a nutriculture system.

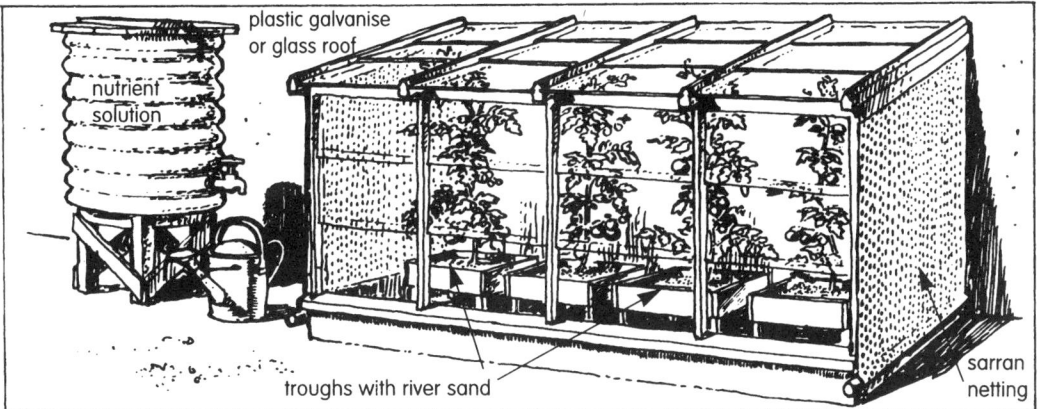

Nutriculture is the system of cultivating plants in soil-less materials or media such as water (hydroponics), gravel (gravelculture), sand (sandculture) and air (aeroponics). These media or materials supply very small quantities of the essential plant nutrients. Plants therefore must be supplied with all of the essential nutrients by means of fertilisers dissolved in water in a carefully prepared nutrient solution.

Hydroponics is one form of nutriculture. Crops are cultivated in a water medium using a technique known as NFT – nutrient film technique, as shown in the diagrams below:

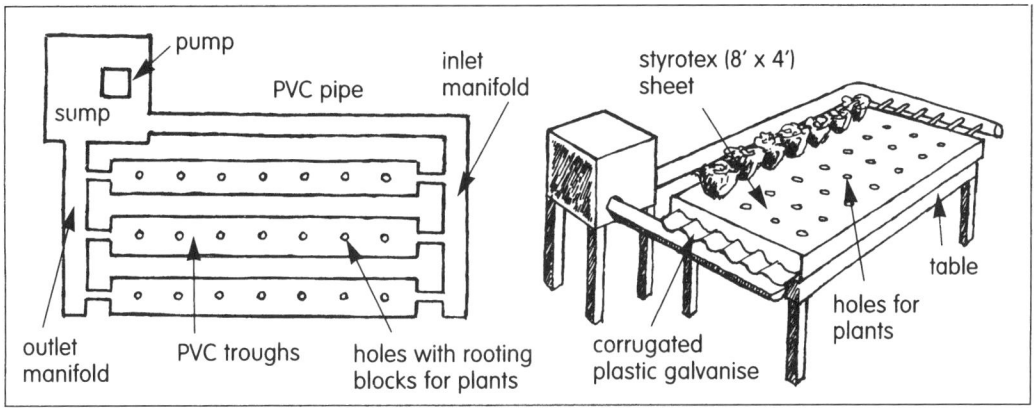

Nutriculture, including hydroponics, is recommended especially for areas where the land *in situ* is not suitable for agriculture. Some other features and advantages are as follows:

— It can be used for cultivating certain crops all year round, for example tomatoes, which normally do not thrive well in the rainy season.
— Since the area is covered, nutrients are not lost through leaching by rain.
— Sunlight passes through the transparent roof to the plants below.
— Because the total area is small, it is much easier to control weeds as well as other pests and diseases.
— Watering can be controlled because of the rainproof roof.
— Plants utilise and respond to the nutrients more quickly because the nutrients are dissolved in water and applied in a liquid form.
— Crop yield is generally higher when compared to the crop yield from an open field of the same area.
— Generally, the crops themselves are of a better quality.

5 Grow-boxes

Grow-boxes are rectangular frames constructed on the surface of the land and filled with a rooting medium for the cultivation of crops. Frames can be constructed using lumber, bamboo, bricks or galvanise and may vary in size depending on the available land area. A convenient size, however, is 10 m long, 1.5 m wide and 25 cm deep as shown in the diagram below:

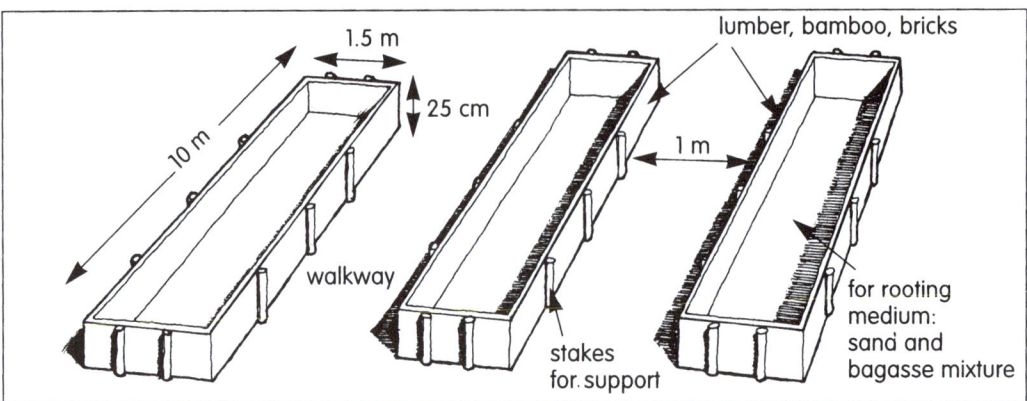

What materials are used for preparing the rooting medium for filling the grow-boxes?
Some materials which are commonly used are as follows:
— 2 parts of sharp sand, or river sand, or plastering sand, or sea sand (washed);
— 3 parts of bagasse, or coconut fibre-bast, or sawdust, or coffee hull, or rice husk (chaff or grainless paddy after winnowing harvested grains).
A combination of materials may also be used and must be properly mixed to produce a rooting medium which is fluffy, porous, loose, easy to till and manage and *not* toxic to plants.

Remember that plants cultivated in grow-boxes need the same kind of care and management as crops grown in garden beds.

Now do **Activity 7** in your Workbook.

6 Seedboxes, seedbeds and speedling trays

(a) Seedboxes

Have you ever seen a seedbox before? What is another name for a seedbox? What is the main purpose of a seedbox?
A seedbox is also known as a nursery box because it is normally used for growing nursery plants or seedlings.
Here is a diagram of a seedbox. Observe it carefully and name its features.

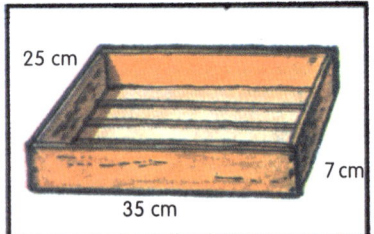

1. Made of lumber 1 cm thick.
2. It is 35 cm long, 25 cm wide and 7 cm deep.
3. Can accommodate a fairly large number of seedlings.
4. Light and convenient to handle.
5. Very suitable for growing seedlings for small gardens.
6. It has slits (5 mm wide) at its base for drainage.
7. It is pleasant in appearance.
8. It is simple to construct.

Now name the tools and equipment you will need for constructing a seedbox. The diagrams below will help you.

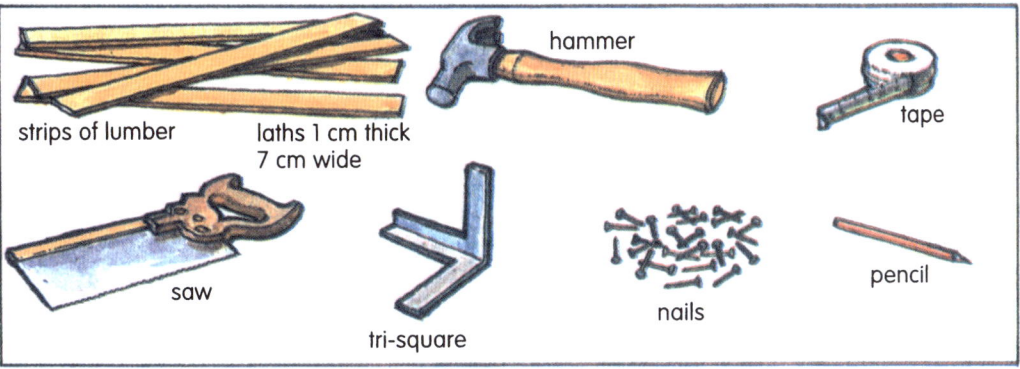

In the Caribbean, other containers are used for growing nurseries or seedlings.
Find out some of their names from the diagrams below:

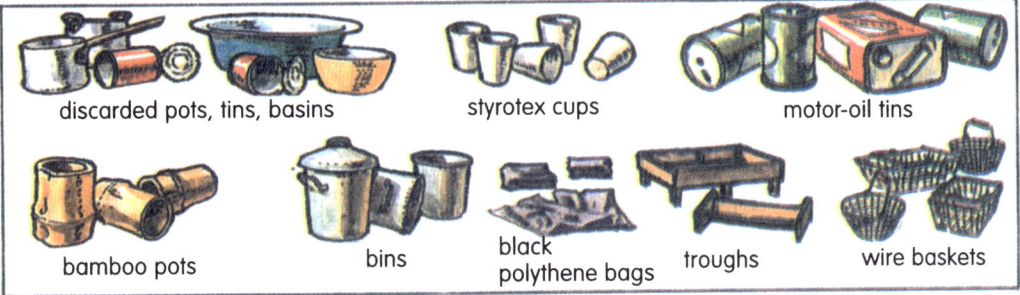

Do you remember the steps in preparing a seedbox? The diagrams below will help to remind you of the procedures.

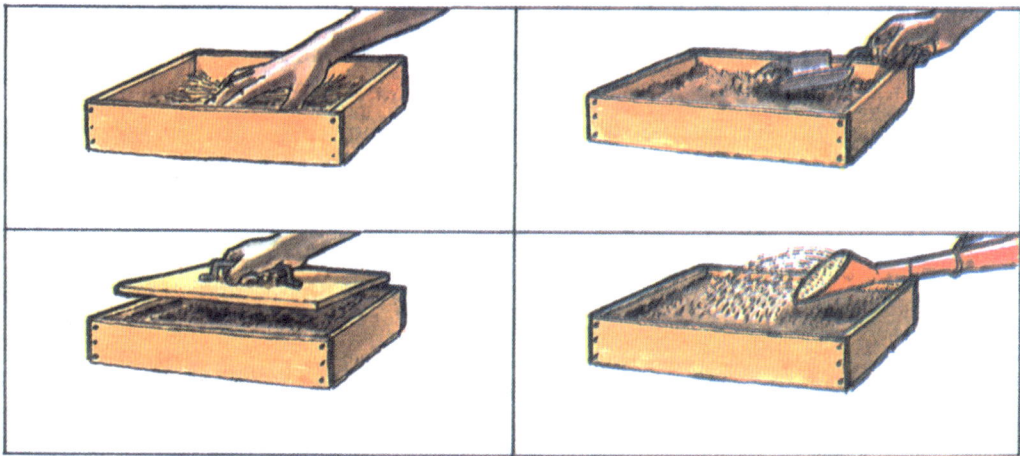

1 Place layer of straw or dry grass on the base inside.
2 Fill with potting soil to height of 1 cm from the top. Why?
3 Use pressboard to level off and to firm the soil gently.
4 Water the soil in the seedbox.

(b) Seedbeds

Look at the diagram of the seedbed below and name its features.

1. Constructed in the garden itself.
2. Suitable size 3 m long, 1 m wide.
3. Cambered bed is made.
4. Box drains surround it.
5. Can accommodate a large number of seedlings.
6. Specially recommended for large gardens.

Observe its size and rectangular shape. Have you noticed the box drains?
Why is the bed cambered?
It is cambered to enable surface water to drain off quickly.
Observe that the seedbed is fairly large and can therefore accommodate a large number of seedlings.
Therefore the seedbed is recommended for growing seedlings for large gardens or field plots.
Explain why a seedbed should be constructed in the garden or field plot.
What is the purpose or function of a seedbed?
How do you go about making a seedbed? Follow these steps:

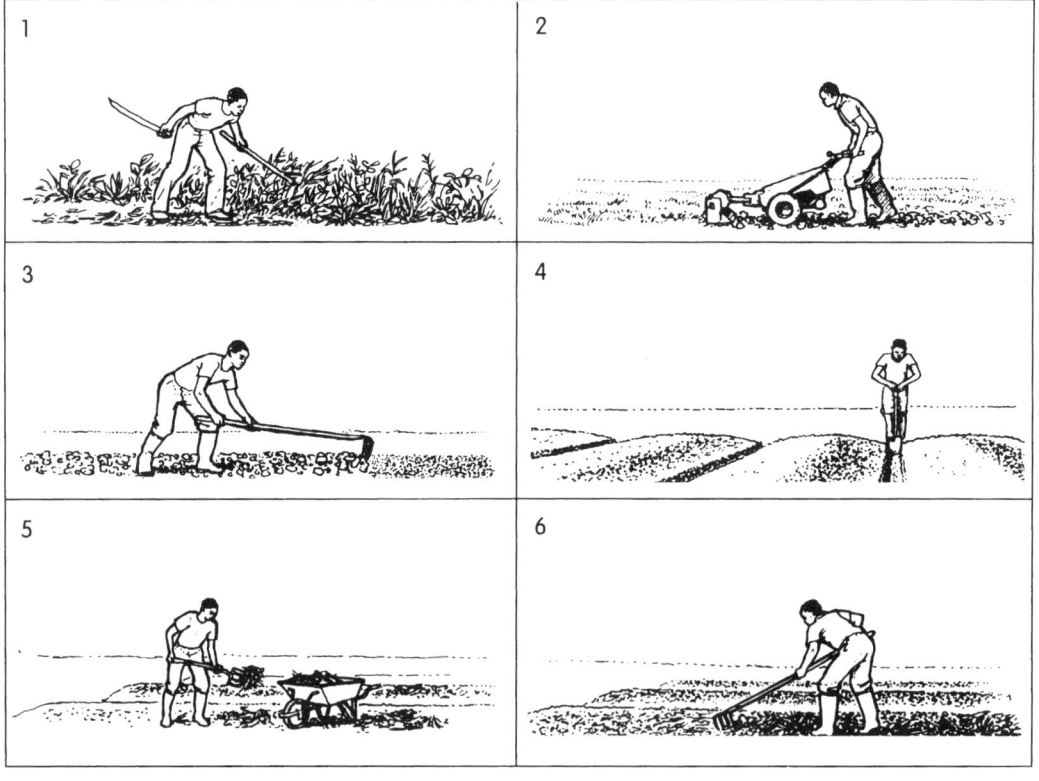

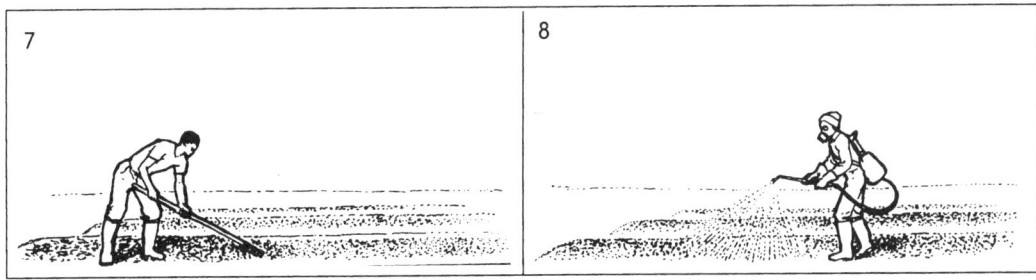

1 Clear the area of all grass and bush.
2 Plough the area using a hand tractor or fork.
3 Refine the soil, using a rotavator, hoe or rake, to a very fine tilth.
4 Dig box drains then camber the bed.
5 Apply manure to the area:
   (i) pen or compost manure (2–4 cm thick)
   (ii) NPK (10:15:10) fertiliser – about 30 gms per square metre (30 gm/m^2).
6 Mix or incorporate the manure within the first 4 to 8 cm of soil.
7 Use a rake to level off the surface and to remove large pieces of organic matter.
8 Spray the seedbed with a mixture of soil insecticide and fungicide (see potting soil).

(c) Speedling trays
Look at the diagrams of the speedling trays below. Observe them carefully and describe the features.

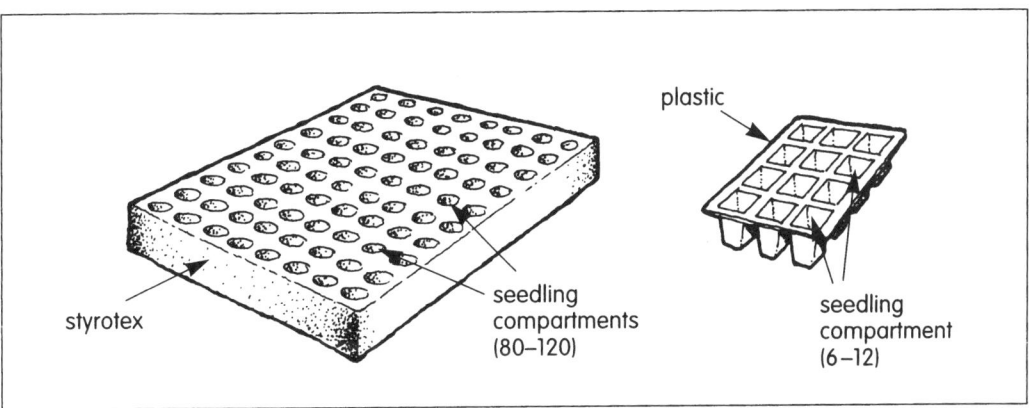

1 Made of styrotex or plastic.
2 Light, easy to handle and carry around.
3 Has several compartments for seedlings.
4 Small speedling tray can hold 6–12 seedlings.
5 Large speedling trays can hold 80–120 seedlings.
6 Speedling trays may be filled with potting soil or promix.
7 One seed is normally sowed in each compartment.
8 Thinning out or pricking off is not normally required.
9 Each seedling is easily uprooted for transplanting.
10 Speedling trays can be reused over a long period.

Speedling trays enable us to do certain nursery operations more speedily. We save time and labour because pricking off or thinning out of seedlings is not required. This is why they are called speedling trays.

Now turn to your Workbook and do **Activity 8**.

Unit 2 Selection and management of planting materials

1 Plant propagation

There are two main methods of propagating plants: natural methods and artificial methods. Let us now study each method in more detail.

(a) Natural methods of plant propagation

Can you name natural methods of propagating or obtaining new plants?
Study the table below to find out the main natural methods of plant propagation as well as local examples of plants propagated by each natural method.

Name of natural method   *Local examples*

Name of natural method	Local examples
1 seeds	corn, pigeon pea, tomato, coconut
2 runners	savannah grass, pumpkin, strawberry
3 suckers	banana, chive, aloe, pineapple
4 (stem) tubers	yam, cush-cush, dahlia (root tubers such as cassava and carrot do not grow to form new plants)
5 rhizomes	arrowroot, ginger, saffron, topi tambu
6 corms	dasheen, eddo, tannia, gladiolus

(b) Artificial methods of plant propagation

The table below gives you the names of the main artificial methods of plant propagation as well as local examples of plants propagated by each method.

Name of artificial method	Local examples
1 Cuttings (stem, leaf)	sugar cane (stem), sweet potato (stem), snake plant or mother-in-law tongue (leaf), wonder of the world or leaf of life (leaf), cocoa (stem)
2 Layering	rose, croton, hibiscus, lime
3 Budding	citrus (orange, grapefruit), rose
4 Grafting	avocado, mango, sapodilla

There are several techniques or methods of propagating plants by cuttings, layering, budding and grafting. Observe carefully as your teacher demonstrates how to propagate plants by each artificial method.

Can you describe one method of layering, budding, grafting and propagation by cutting?
The diagrams below will help you.

(i) Rooting a stem cutting in potting soil

1. Fill container (pot, polythene bag) with potting soil and water the soil.
2. Select and cut suitable stem cutting from croton plant very early in the morning.
3. Use a round stick and make a hole in the potting soil.
4. Dip the base of the cutting into a rooting powder.
5. Place cutting in the hole and gently compress or firm the soil around it.
6. Water the soil and place the container in a cool area for rooting to take place.

(ii) Air-layering or marcotting

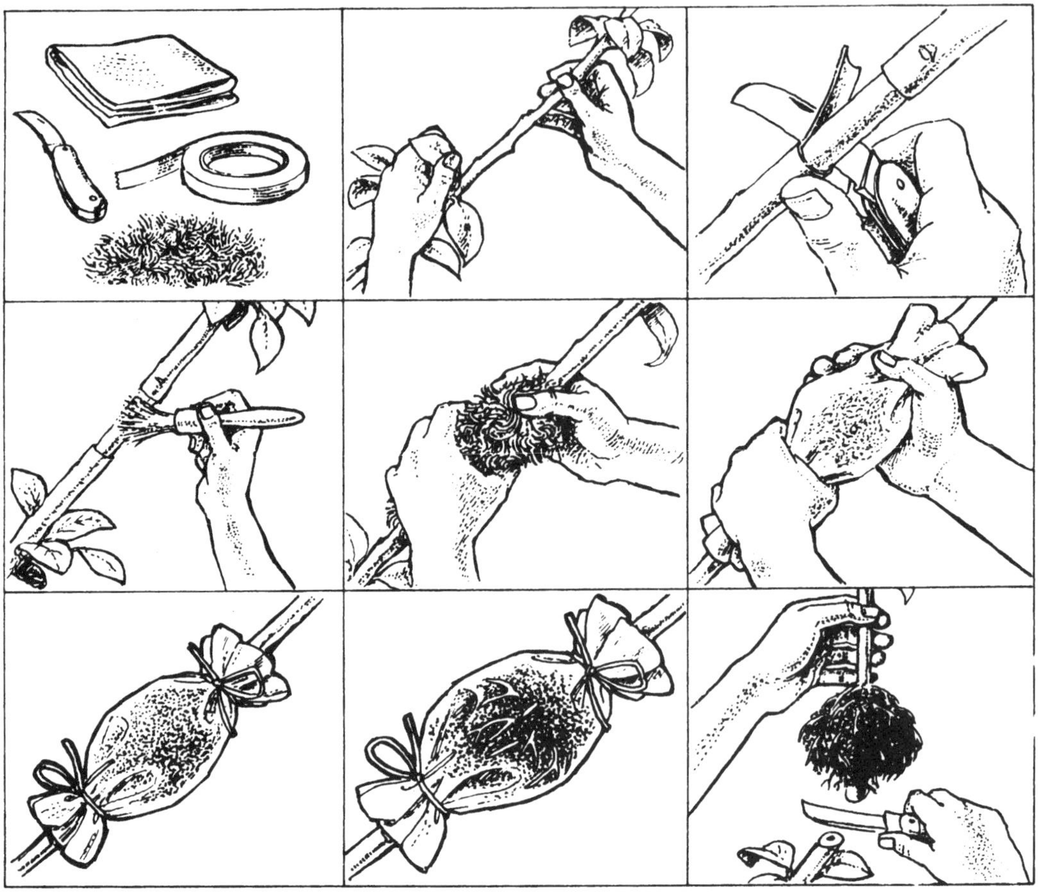

1. Obtain materials: knife, polythene tape or strips, moist moss, clear polythene sheet.
2. Select stem and remove a few leaves in the area to be layered.
3. Cut or girdle stem and remove bark.
4. Apply rooting powder (hormone).
5. Apply moist moss or coconut fibre.
6. Place clear polythene sheet around moss.
7. Tie with twine.
8. Observe! Roots should appear in 14–21 days.
9. Remove from parent plant and pot.

(iii) Inverted 'T' budding in citrus

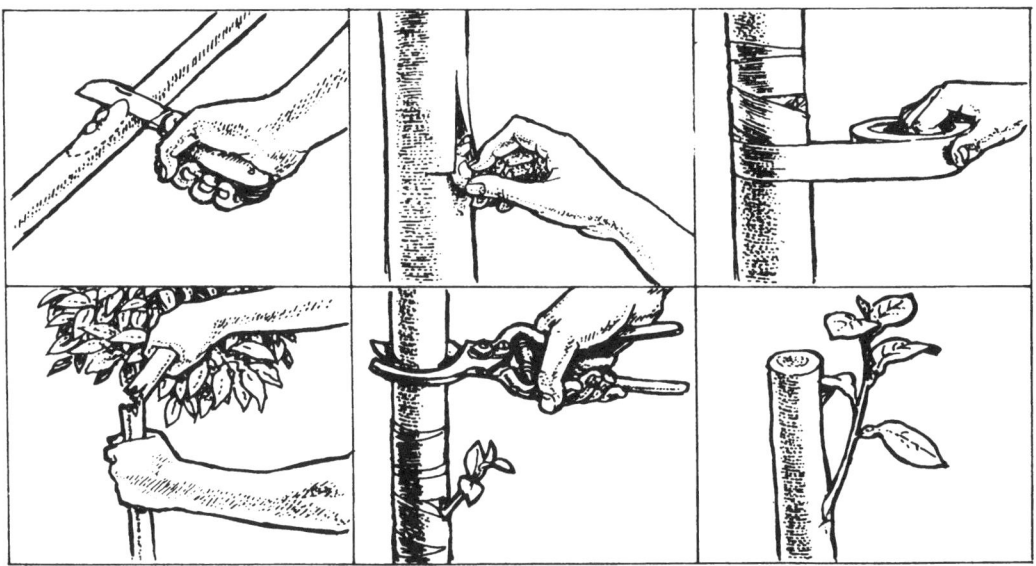

1 Obtain materials: budding knife, polythene tape or strips, cleopatra mandarin plant (called 'stock'), orange budwood (called 'scion').
2 Make incision on stock.
3 Lift bark.
4 Cut out bud (scion).
5 Insert bud or scion in stock.
6 Tape scion onto stock.
7 Break top of stock plant.
8 Cut off top of stock plant.
9 Observe: budded citrus plant.

(iv) Side graft in mango

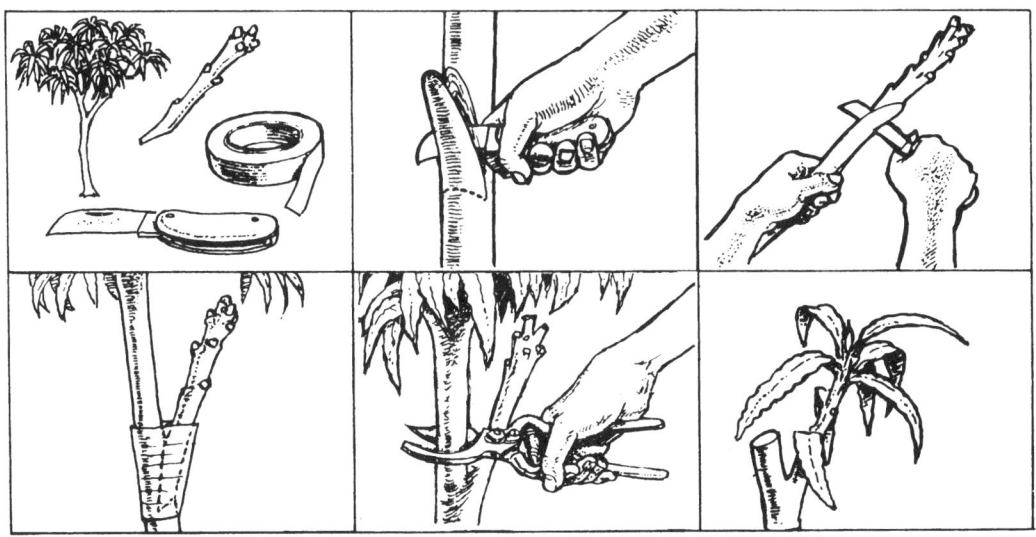

1 Obtain materials: budding knife, polythene tape or strips, Rose mango plant (stock), Julie mango budwood (scion).
2 Make incision in Rose mango stock.

3 Prepare Julie mango scion.
4 Insert scion in stock and tape carefully.
5 Break and cut off top of stock.
6 Observe: grafted Julie mango plant.

Now do **Activity 9** in your Workbook.

2 Planting materials
(a) Types of planting materials

The various methods of plant propagation enable us to obtain different types of planting materials as shown in the chart below.

Find out the names of crop plants which are planted directly in the field plots. Name plants which require nursery treatment and transplanting.

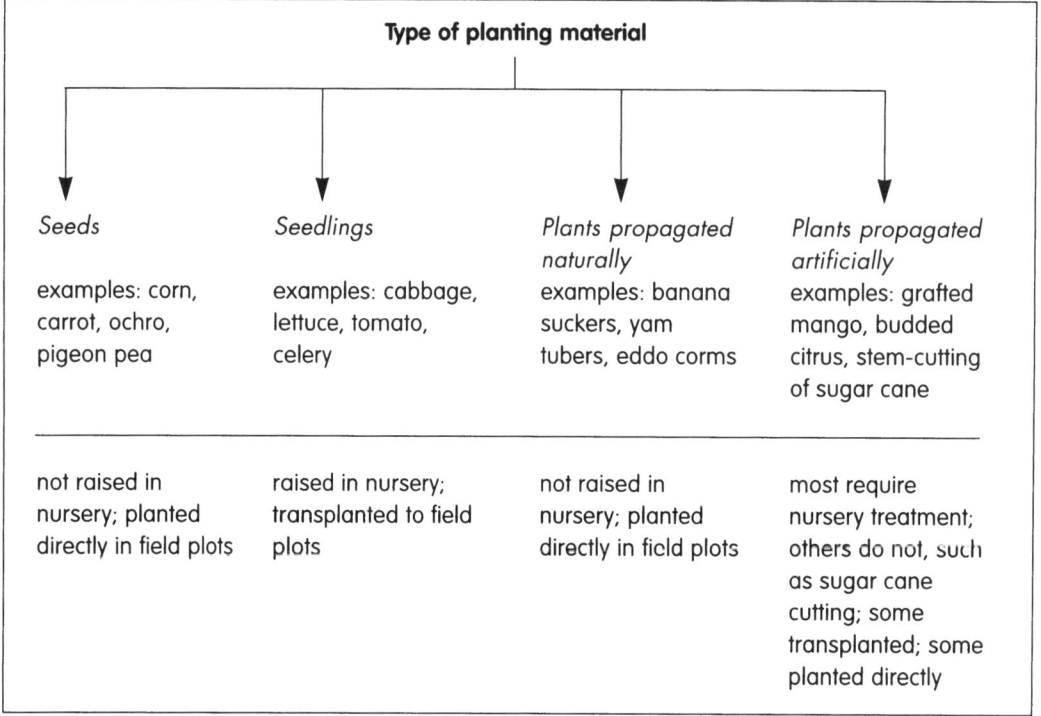

(b) Selection of planting materials

What are some major points to bear in mind when selecting planting materials? Check the list below to find out:
— select planting materials which are high yielding, for example, hybrid seeds, improved varieties.
— choose planting materials which will bear good quality produce so that it will be marketed.
— select varieties of the particular crop which are best suited for the season (dry season variety or wet season variety, year-round varieties).
— choose planting materials which are free from pest and disease attack.
— select planting materials which are viable (that is, will grow).
— it is preferable to select planting materials which have been treated; if not, treat them before or at planting time.

Now turn to your Workbook and do **Activity 10**.

(c) Nursery techniques or practices
(i) Sowing seeds in seedboxes, speedling trays and seedbeds
Here are some methods of sowing seeds in seedboxes, speedling trays, seedbeds and other containers:

1 Scattering or broadcasting seeds.
2 Sowing seeds in drills.
3 Mixing seeds with water in watering can, stirring properly and watering the seedbed: for example tobacco seeds.
4 Planting seeds singly in containers: for example seedling cocoa.

Carefully observe and assist your teacher as he shows you how to sow seeds.

Do you know the three necessary requirements for the germination of seeds? What are they?
They are, of course, *air*, *warmth* and *moisture*.

Watering provides *moisture* to the seeds.
Observe that the potting soil is loose, therefore *air* enters the soil and gets to the seeds.
Notice that your teacher covered the seeds with a thin layer of fine, potting soil. Why? It is because the covering of soil gives *warmth* to the seeds.
Covering the seedbox or seedbed with coconut palm leaves, or with sarran netting, also provides warmth.
However, the covering must be removed as soon as germination is visible. Why?
Find out by leaving one seedbox covered.
Observe that the seedlings are whitish, long, lanky and very weak. This feature or phenomenon is known as *etiolation*.

(ii) Thinning out or pricking off seedlings
Now check the diagrams below to find out how to thin out or prick off seedlings.

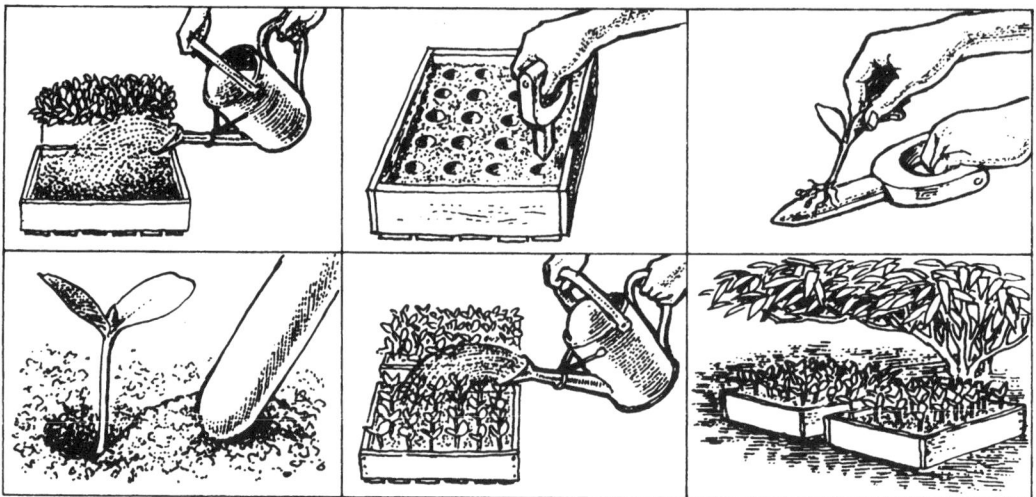

1. Water seedlings and soil in newly prepared nursery box.
2. Use a *dibber* and make holes in rows with 5 cm spaces around in the prepared seedbox.
3. With thumb and index finger hold one leaf of each seedling and lift off with dibber.
4. Place root of seedling carefully in each hole and use dibber gently to firm soil around it.
5. Water both the seedbox and the newly pricked off seedlings, after thinning out.
6. Protect newly thinned out seedlings from direct sunshine and rain.

Pricking off or thinning out is also done on seedbeds. Your teacher will demonstrate to you and also give you the opportunity to prick off seedlings.

Can you explain why seedlings are thinned out?
Here ae some reasons:
— to allow each seedling more space and root room;
— to reduce the competition for light, space, water and nutrients;
— to enable each seedling to grow vigorously and healthily;
— to enable each seedling to be dug out with a 'pillon' (ball of soil attached to the roots) when transplanting.

Now do **Activity 11** in your Workbook.

(d) Management or care of young plants
Carefully study the practices outlined below. They will enable you properly to manage or to take care of young plants.

1 Water plants regularly

Preferably, a good watering can should be used.
Avoid using a powerful pressure hose. Why?

2 Uproot or dig out weeds

Weeds rob the young plants of light, space, water and nutrients. Some also encourage or harbour pests and diseases. Therefore, weeds must be removed regularly.

3 Break up soil surface to aerate the soil

You may observe that the soil surface of nursery containers tends to become compact or crusted. This prevents air and water from getting to the roots of plants. Therefore, the soil surface must be broken up with a dibber.

4 Apply manure to plants

After breaking up the soil surface, fine pen manure can carefully be spread on the soil. Liquid manure can also be added by dissolving about 10 gm urea in 4.5 l of water and watering the young plants.

5 Control pests and diseases

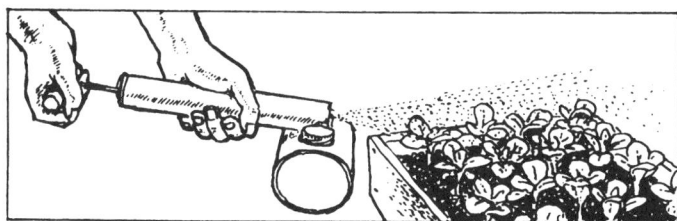

Common pests in young plants are caterpillars, cutting ants (bachacs), mole crickets, leaf miner and slugs. Use a soil insecticide such as Diazinon to control soil pests. Other insecticides such as Malathion and Lannate will control leaf miner and caterpillars.

A common disease in the nursery is 'damping off'. It is caused by the Pythium species of fungus and causes the young seedlings to topple over and 'melt'.
Damping off can be controlled by doing the following:
— cutting down or controlling watering;
— aerating the soil;
— spraying plants twice weekly with a fungicide such as Banrot or Captan.

6 Harden young plants

Expose young plants gradually to sunlight until they are receiving full sunlight.
Why? This will help to harden or strengthen the plants so that they will be able to withstand the full sunlight when transplanted.

(e) Procedures for planting out
(i) Direct planting
Do you remember the names of some planting materials which are planted directly in the field or garden plots? Name some of them.

Now study the diagrams below to find out the steps or procedures for planting out materials directly in the field plots:

1 Line up planting materials

This is done so that the plants grow in rows and because it makes the garden attractive.

2 Use the correct spacing for the particular crop

Spacing allows each plant enough room to grow and develop so that weeds are smothered when the plants grow up. It also makes maximum or efficient use of the land.

3 Dig holes to accommodate the planting material

Digging holes is to make sure the material is beneath the soil for anchorage.

4 Incorporate some pen or compost manure

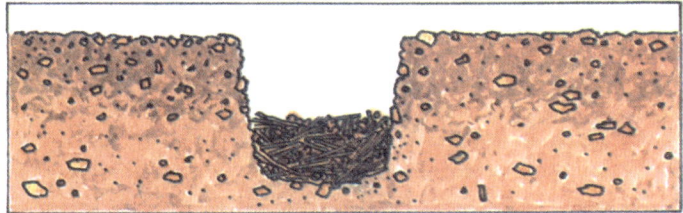

Manure supplies nutrients and holds moisture.

5 Avoid planting the materials deeply

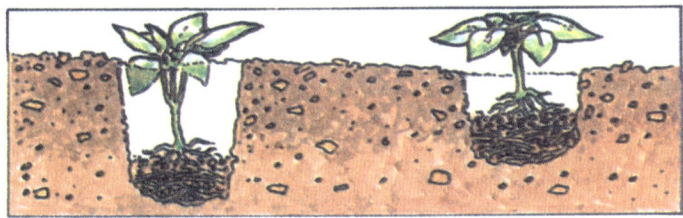

If materials are too deep they may fail to germinate. Also some materials may rot.

6 Water or irrigate after planting

Watering prevents wilting or drying out of the materials and supplies moisture for germination.

(ii) Transplanting (seedlings and young plants)
Can you describe some steps you should take when transplanting seedlings?
At what time of the day should transplanting be done? Why?
Transplanting should be done very early in the morning, late in the evening or when the weather is cloudy. The reasons are:
— to allow the seedlings enough time to recover from the shock of transplanting;
— to protect the seedlings from the direct, scorching effect of the sun whilst transplanting;
— to reduce the wilting effect which occurs at transplanting time.

Follow procedures 1 to 4 as outlined for materials which are planted directly in the field plots:
1 Line up planting materials
2 Use the correct spacing for the particular crop
3 Dig holes to accommodate the planting material
4 Incorporate some pen or compost manure in each hole
In addition, these steps should be taken:

5 *Wet the soil in the nursery box, speedling tray or container*

The water enables the soil particles to cohere, or stick together, and reduces wilting.

6 *Remove each seedling with a 'pillon' using a trowel or uproot seedling from speedling tray*

This avoids damage to the young roots and reduces the shock of transplanting.

7 *Avoid planting too deeply and gently firm soil around plant*

This prevents stem and root damage and keeps the young plant upright.

8 *Water seedlings after planting*

The water reduces wilting and further firms or settles the soil. It also brings the soil into closer contact with the roots.

Now turn to your Workbook and do **Activity 12**.

Unit 3 **Field management of crops**

1 Cultural practices

The success of any business depends on proper management.
Farming is a business. Therefore, in order for farmers to grow crops successfully they most carry out proper management practices.
In agricultural terms, these management practices are known as *cultural practices*.

Can you name some cultural practices that farmers adopt or use? Listed below are some major cultural practices which successful farmers use.
Study each one carefully. Your teacher will demonstrate to you how each cultural practice is actually done.

(a) Moulding

Moulding is also referred to as 'earthing-up'. Observe that small mounds are made around the base of plants.
Advantages:
— keeps the plants upright;
— prevents surface water from settling around the base of plants;
— encourages the growth of adventitious roots, as in tomato plants.
Disadvantage:
— when moulding is done in the dry season it may prevent water from getting to the root zone.

(b) Staking

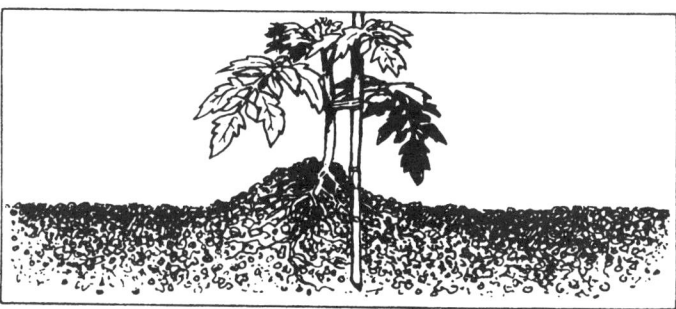

Staking involves the placement of stakes in an upright position beside plants. Observe that the basal portion of each stake is firmly buried or stuck in the soil. Also observe that it may be necessary to tie the stems of some plants to the stakes, for example, tomato.

Advantages:
— supports weak stems or vines;
— keeps fruits off the wet ground.
Disadvantage:
— staking may be a problem when land clearing is done.

(c) Pruning

Pruning means cutting off diseased, damaged and unwanted parts of crop plants.
Advantages:
— keeps plants healthy;
— maintains plants in a proper shape;
— encourages growth of a single stem, for example tomato;
— obtains larger fruits.
Disadvantage:
— pruning can spread diseases from diseased plants to healthy plants. Can you explain how?

(d) Inter-row cultivation

Inter-row cultivation is the digging up or loosening of the soil between rows of crop plants.
Advantages:
— controls weed growth;
— aerates the soil;
— allows water to enter the soil more freely.
Disadvantage:
— can lead to damage of roots, stems and leaves of crop plants if not done properly and at the correct time.

(e) Mulching

Mulching involves the covering of the soil surface, especially in the dry season, with materials such as grass, straw, polythene, styrotex-pellets, stones and dust (obtained through rotavating the soil on the spot). Place mulch such as grass or straw about 10 cm from the stem to a thickness of 10 to 15 cm.
Advantages:
— conserves soil moisture;
— controls weeds.
Disadvantage:
— some mulching materials can harbour pests and diseases, for example, crop residues, weeds.

(f) Crop rotation

Crop rotation is the cultivation of different families of crop plants in succession (one crop followed by another crop) on the same plot of land. An example of a good crop rotation is tomato → beans → lettuce → carrots.
Advantages:
— makes use of nutrients at different levels in the soil (plant shallow-rooted then deep-rooted crops);
— controls pests and diseases;
— improves and maintains the soil in a fertile state (for example, legume crops add nitrogen to the soil).
Disadvantage:
— it is not practicable for long-term or permanent crops.

(g) Inter-cropping

Inter-cropping as a cultural practice involves the cultivation of short-term crops (for example lettuce) amongst other major crops (for example, ochro).
Advantages:
— more efficient land use;
— obtains more farm income or profits;
— controls weed growth.
Disadvantage:
— the inter-crop may encourage and harbour pests and diseases which can later attack the main crop.

(h) Cover-cropping

Cover-cropping is the cultivation of crops which will give quick coverage to the soil surface, for example pumpkin, cucumber, kudzu.

Advantages:
— controls soil erosion;
— conserves soil moisture;
— controls weed growth;
— adds organic matter to the soil.

Disadvantage:
— can encourage and harbour soil pests.

Now do **Activity 13** in your Workbook.

2 Drainage and irrigation

(a) Drainage

Look back to pages 13 and 14 of this textbook and revise what you have already studied about drainage. Now can you answer these questions?
— what is drainage?
— what tools and equipment are used for drainage?
— name and describe some kinds of drains.
— state the importance or benefits of drainage.

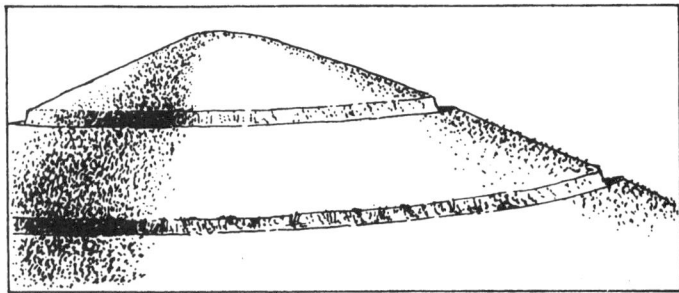

Now look at the drains in the diagram above.
Observe that the drains are dug across the slope along the *contour*, that is, on land which is of the same height above sea level.
What is the name of this type of drain?
These drains are known as *contour drains*, of course.
Observe that contour drains are used on *hilly* land.

Explain what will happen if the drains are dug down the slope.
Soil erosion will take place if the drains are dug down the slope.
Contour drains, therefore, help to prevent soil erosion.

How can you tell when drainage is necessary in a field?
Find out from the pictures below:

1. Water collects on the field plots or beds.
2. The soil is always moist and soggy.
3. Ground is very moist and leaves of crop plants show signs of yellowing.
4. Crop plants may wilt because of root damage.

Can you explain how drainage might have disadvantages for crop cultivation?

Rice (paddy) is a water-loving crop and so it thrives best in water.
Drainage of paddy fields, therefore, will result in poor crop yields.

In the dry season rainfall is infrequent. However, the small amount of water from rain is essential for crop cultivation.
Surface water from rain and irrigation is therefore allowed to soak or penetrate into the soil to the roots of crop plants. For this reason garden beds are made flat-topped and holes for planting are bowl- or basin-shaped.

As you can see, the problem here is one of water conservation rather than drainage.
Drainage in the dry season, therefore, can have disadvantages for crop plants.

(b) Irrigation
What is irrigation?
Irrigation is the process or means of supplying water to crop plants.

Of what importance is irrigation to crops?

Irrigation is important for
— dissolving mineral salts (nutrients) in the soil so that the roots of crop plants may absorb them;
— maintaining a desirable amount of water in the soil so that crop plants will not wilt and die.

Can you name some systems of irrigation which are commonly used in the Caribbean?
Study the diagrams below to find out.

1 Manual irrigation   2 Furrow irrigation   3 Sprinkler irrigation

Which of these irrigation systems are best suited for:
(i)   your school garden?
(ii) a pasture?
(iii) a large corn field?
(iv) a lawn?

How can you tell when irrigation is necessary? Check the diagrams below.

crop plants wilt   soil becomes dry

Irrigation is normally needed
(i) especially in the dry season
(ii) after seeds are planted
(iii) after seedlings are transplanted
(iv) throughout the life of crop plants.
Here now are some ways in which irrigation can be disadvantageous or harmful to crop plants.
— when an excessive amount of water is supplied to crops and causes water-logging;
— when there is so much surface run-off that soil erosion occurs;
— when irrigation water is applied with too much force and knocks off flowers and causes damage to crop plants;
— when the irrigation water contains substances which are toxic or poisonous to crop plants, for example, salt, weedicide.

Now do **Activity 14** in your Workbook.

3 Pests and diseases

(a) Pests

How would you describe a pest?
A pest is any organism which causes harm or injury to crop plants.
Pests may be of different types as shown in the chart below:

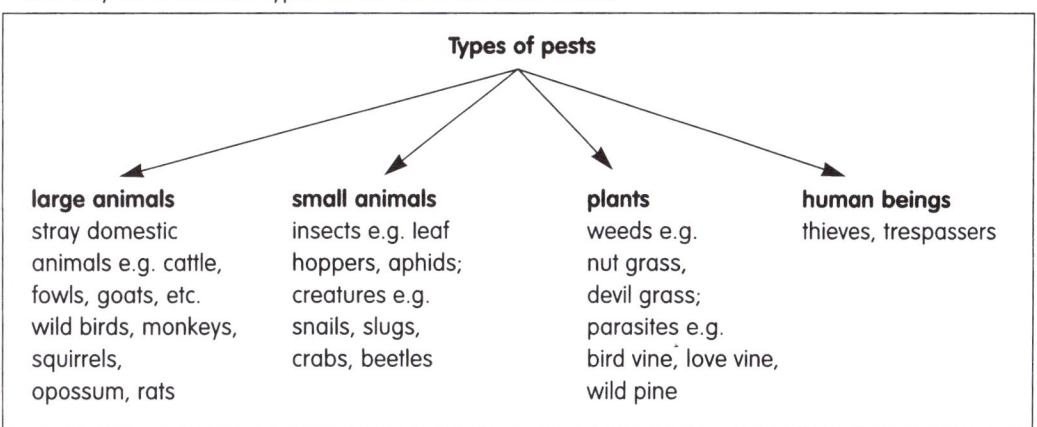

large animals	**small animals**	**plants**	**human beings**
stray domestic animals e.g. cattle, fowls, goats, etc. wild birds, monkeys, squirrels, opossum, rats	insects e.g. leaf hoppers, aphids; creatures e.g. snails, slugs, crabs, beetles	weeds e.g. nut grass, devil grass; parasites e.g. bird vine, love vine, wild pine	thieves, trespassers

Now point out and name the pests in the diagram below:
State to which group each pest belongs.

monkey	caterpillar	wild bird	slug
mealy bug	squirrel	aphids	crab	snail
leaf hopper	nut grass	thieves	weevil
wild pine	love vine (dodder)	scale insect	bird vine (mistletoe)

41

Can you name some pests and describe the kind of damage that they do to crop plants?
The diagrams below will help you to identify some pests and pest attacks in crops.

1 caterpillar attack	 2 flea beetle attack	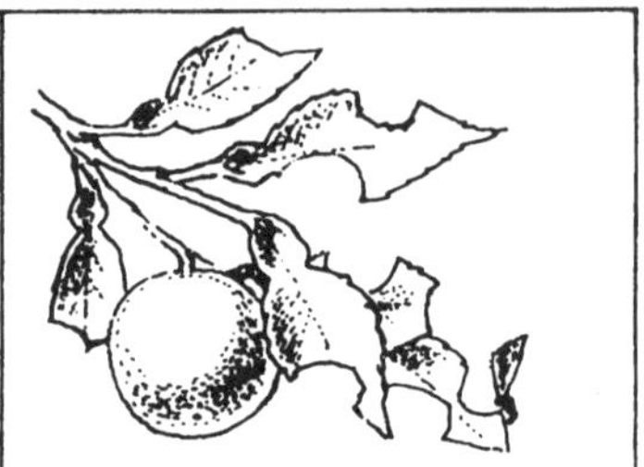 3 bachac attack
 4 damage by monkeys	 5 damage by squirrels	 6 damage by birds
 7 aphid attack	 8 bird vine attack	 9 mole cricket attack

Now, make regular visits to your school and home gardens and try to identify more pests and pest attacks in crop plants.

Now turn to your Workbook and do **Activity 15**.

(b) Diseases

Explain what is meant by the word 'disease'.
Disease is the term used to describe anything which makes a plant unhealthy or abnormal.
What causes diseases in crop plants?
Diseases in crop plants are caused by micro-organisms or agents such as bacteria, fungi, viruses, nematodes and mycoplasma.

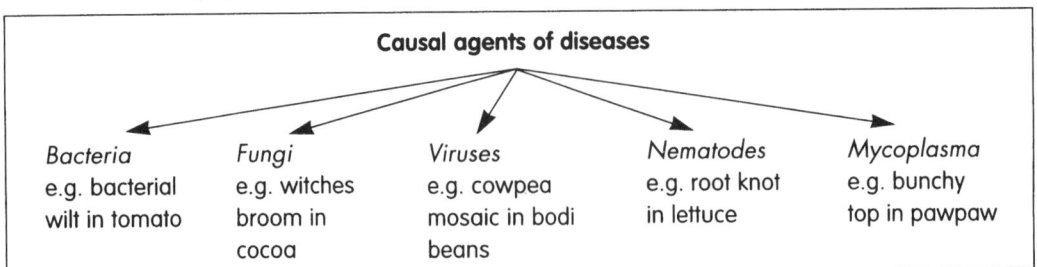

Causal agents of diseases

Bacteria	Fungi	Viruses	Nematodes	Mycoplasma
e.g. bacterial wilt in tomato	e.g. witches broom in cocoa	e.g. cowpea mosaic in bodi beans	e.g. root knot in lettuce	e.g. bunchy top in pawpaw

In each of the groups of causal agents, there are several sub-groups or species.
Certain species of each group are responsible for causing specific plant diseases.
It must be remembered, however, that not all species of bacteria, fungi and viruses are harmful to crop plants.
In fact, some are very useful in agriculture, for example:
— certain species of bacteria increase the nitrogen content of soils;
— antibiotics are obtained from certain species of fungi;
— vaccines are made with certain species of viruses.

Now, can you explain how the causal agents of diseases are spread to crop plants?
Generally, agents of disease are spread or dispersed by:
1 *wind* – spreads bacteria, viruses and fungal spores;
2 *water* – splashing raindrops spread bacteria, viruses and fungal spores; soil water spreads nematodes;
3 *animals* (including birds, insects and other creatures) – carry or transmit agents of disease from diseased plants to healthy plants;
4 *human beings* – transmit agents of disease from diseased plants to healthy plants by handling, pruning, budding and grafting.

Living organisms which carry or transmit causal agents or disease from diseased to healthy plants are known as *vectors*.
Vectors are infected, that is, they actually carry the agents of disease within their bodies.
Therefore, every time a vector (such as an aphid) bites, pierces and sucks, or lays eggs on a healthy plant it is actually depositing agents of disease.

Some vectors are responsible for the spread of specific diseases as shown in the table:

Vector	Causal agent transmitted	Disease caused	Crop attacked
Aphid	virus	tobacco mosaic virus disease	tobacco, tomato
Leaf hoppers	mycoplasma	bunchy top disease	pawpaw
Palm weevil	nematode	red ring disease	coconut

Here now are some common diseases in crop plants.
In each case, note carefully the causal agent and the crop which is attacked.

Disease	Name	Causal agent	Crop attacked
	bacterial wilt	bacteria	tomato
	anthracnose	fungus	melongene
	leaf spots	fungus, bacteria	beans
	powdery mildew	fungus	cucumber
	root knot	nematode	tomato
	rust	fungus	pigeon pea
	black pod	fungus	cocoa
	internal rot	nematode	yam
	cowpea mosaic	virus	bodi beans

So far we have only looked at diseases which are caused by causal agents such as bacteria, fungi, nematodes, viruses and mycoplasma.

However, diseases in crop plants are also caused by the following:

1 Lack of certain nutrients in the soil, for example, magnesium deficiency; certain specific symptoms may show up in the leaves of plants.

2 Infrequent watering; may cause blossom end rot in tomato, melon and pumpkin.

3 Inadequate drainage; may cause root damage, yellowing and wilting of crop plants.

Now turn to your Workbook and do **Activity 16**.

(c) Methods of controlling pests and diseases

Pests and diseases in crops result in
— poor yields or harvests
— poor quality produce
— severe loss in profits to farmers.

To avoid these losses we must use effective measures for controlling pests and diseases.

Can you name methods which are used for controlling pests and diseases?
Check the diagrams below to find out.

1 Manual method

This method includes hand picking of insect pests, pruning off plant pests and diseased parts of plants, and trapping or shooting animal pests such as squirrels and opossums.

2 Mechanical method

Machines may be used for controlling pests and diseases: tractors with boom-sprayers, light aeroplanes with sprayers and mechanical light traps for trapping insect pests.

3 Chemical method

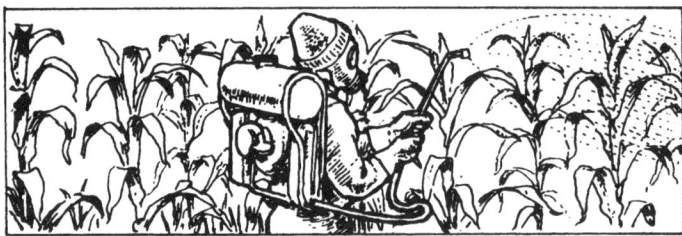

Different types of chemicals may be used for controlling pests and diseases.
Some chemicals which are commonly used include insecticides, fungicides and nematicides.
Can you name an example of each group?

4 Biological method

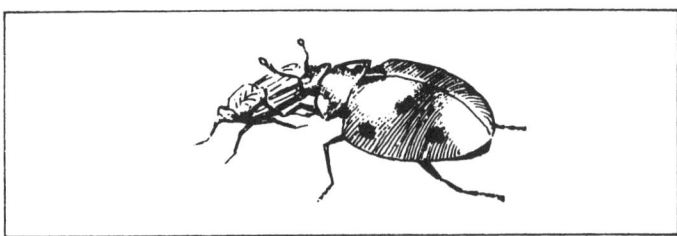

Can you explain what is meant by biological control?
It is a method which involves the use of live animals (including birds and insects) for controlling other harmful pests, for example,
— frogs eat insect pests;
— birds eat insect pests;
— ladybird beetles eat scale insects and mealy bugs.

5 Legal method

Laws are made to prohibit or prevent the importation of certain plant materials into our country. Only disease-free materials are allowed into our country. To be sure of this, imported plants are kept in *quarantine* for a number of months for observation before being released to importers.

6 Cultural methods

Cultural practices for controlling pests and diseases include, for example,
— use of crop varieties which are resistant to certain diseases;
— liming of soils;
— use of fertilisers;
— flooding to control soil pests.

In order to control effectively pests and diseases in crop plants, we must make daily observations in the field to determine what pests and diseases are present. Following these observations, we can then decide what method of control we should use.

Farmers abuse chemicals for controlling pests and diseases too often. We must remember that chemical control or 'spraying' is only one method of controlling pests and diseases. It is advisable to use a combination of the different methods of control, for example manual, cultural, chemical, biological and legal. This is known as an integrated pest management system.

When using chemicals we must take certain precautions. Why? It is because all chemicals are dangerous to human and animal life. What precautions should we take when using chemicals or pesticides? The procedure below should be followed:

1. Choose the correct chemical, that is, use an insecticide for killing insects, a fungicide for controlling fungi and so on.
2. Always read the label carefully.
3. Use the correct dosage as recommended.
4. Wear protective clothing when handling chemicals. that is, long-sleeved shirt, long trousers, gloves, rubber boots, respirator and hat.
5. Do not smoke, eat or drink whilst spraying.
6. Spray very early in the morning or late in the evening when the windspeed is very low.
7. Do not use chemicals that will destroy bees. Why?
8. Avoid washing spray cans or dumping unused chemicals in rivers, drains or ponds. Why? It is because chemicals not only pollute the environment but may also kill or poison fish.
9. After spraying, remove and wash clothing.
10. Bathe and put on clean clothes.
11. Consult a doctor if you feel sick after spraying.
12. Store all chemicals and spraying equipment away from children and animals (including pets) under lock and key.

Now, with the aid and guidance of your teacher make a list of some common pesticides and state the purpose for which they are used in the garden. Here is an example:

Name of chemical	Purpose for which it is used in the garden
1 Diazinon (soil insecticide)	for destroying soil pests such as mole crickets, ants, slugs and snails
2 Sevin (leaf insecticide)	for killing leaf-eating pests such as caterpillars and aphids
3 Kocide 101 (fungicide)	for controlling fungal diseases such as leaf spots, anthracnose and so on
4 Furadan (nematicide)	for controlling nematodes
5 Mirex (insecticide)	for destroying bachacs or 'cutting ants'

Now do **Activity 17** in your Workbook.

4 Manuring and fertilising
(a) Nutrients required by plants
In order for crop plants to grow healthily and produce high yields, they must be supplied with adequate amounts of the essential nutrients. Study the chart below:

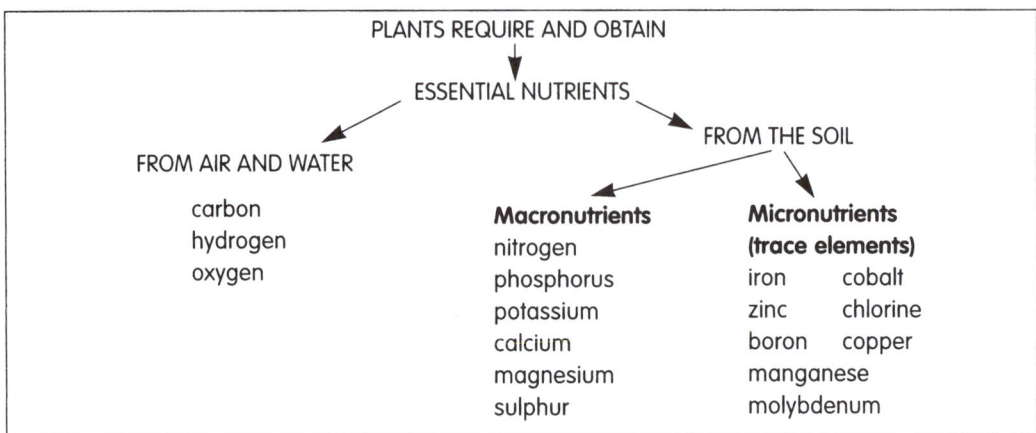

The chart shows that plants require essential nutrients which are obtained from the air, water and soil. Name the nutrients or elements which plants obtain from the soil.

Observe that of the fourteen (14) essential nutrients from the soil, six (6) are **macronutrients** and eight (8) are **micronutrients** or trace elements. Name the various **macro** and **micro** nutrients which are required by plants.

Why are they called macronutrients?
What is another name for micronutrients?
Can you explain why they are called trace elements?
The macronutrients, for example, nitrogen, phosphorus and potassium, are required by plants in fairly large quantities.
Check the table below to find out the role or importance of each of these nutrients in plants.

Macronutrients	Role or importance to plants
nitrogen (N)	for plant growth; for dark green leaves
phosphorus (P)	for proper root growth and development
potassium (K)	for fruit production; for good quality fruits

Micronutrients or trace elements, for example, copper, manganese and boron, are required by plants in small or minute quantities.

(b) Supplying nutrients to plants

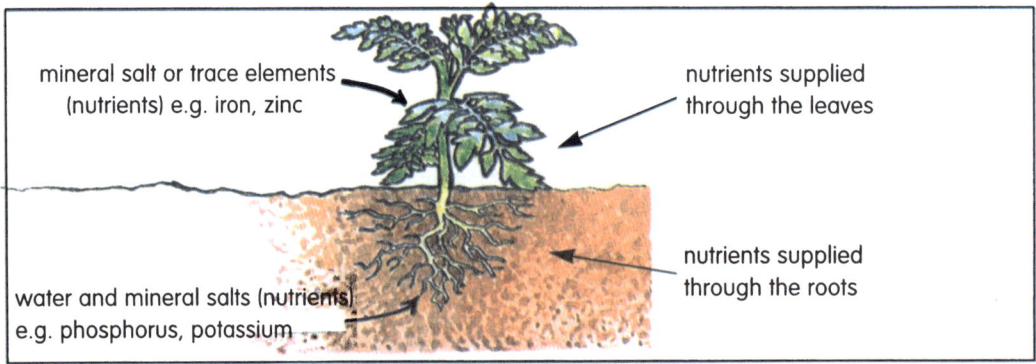

In the previous lesson, we learnt that plants obtain nutrients from the air, water and soil. But how do nutrients enter the plant?

The diagram on page 48 shows that plants obtain:
1. water and mineral salts (nutrients) from the soil through their roots;
2. mineral salts or trace elements (nutrients) through their leaves.

How then must we supply nutrients to plants?
Plants normally obtain most of their nutrients from the soil.
Therefore, we must supply nutrients to the soil so that plants may absorb them through their roots.

However, nutrients can also be supplied to plants through their leaves. Normally, the minor or trace elements are dissolved in water and sprayed on to the leaves of plants, that is by foliar application. This is done especially when the leaves of plants show signs or symptoms that certain trace elements are lacking in the soil.

However, macronutrients can also be supplied through the leaves of plants. This is done by dissolving each of the following fertilisers in 4.5 litres of water and spraying the solution onto the leaves of plants once per week:
1. 10 gms urea – supplies nitrogen;
2. 10 gms nutrex (20:20:20) – supplies nitrogen, phosphorus and potassium.

Now do **Activity 18** in your Workbook.

(c) Manures or nutrient suppliers
Study the chart below and see whether you can define or explain the term 'manure'.

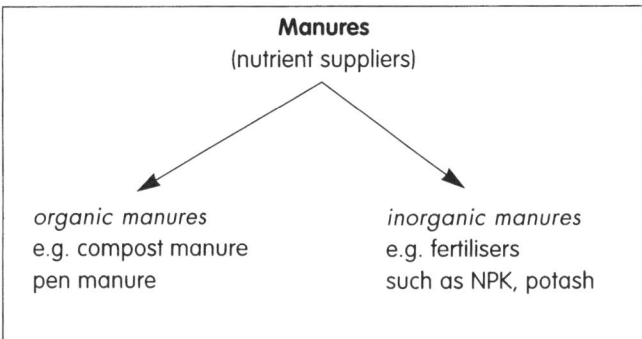

Observe that manures supply nutrients to plants. Notice that there are two main groups of manures. Name them and give an example of each group of manure.

(i) Organic manures
What are organic manures?
Organic manures are obtained from plants and animals.
The names of some types of organic manures are given below.

Types of organic manures

1. pen or farmyard manure

2. liquid manure or slurry

3. compost manure

4. green manure

Now let us study each one in more detail:

Pen or farmyard manure

Carefully observe the pen manure which your teacher has provided for you.
Notice that it is bulky but fairly loose and friable.
From where is it obtained?
It is obtained from animal pens or farms – hence the name pen or farmyard manure.
Can you describe its peculiar smell?

Liquid manure or slurry

As the name suggests, this type of organic manure is in a liquid form.
Look closely at some liquid manure. What is it composed of?
Observe that liquid manure is composed of urine, water and bits of dung or droppings from animals.
It is really the urine and washings from the pens of animals such as cattle and pigs after the straw bedding and bulky droppings have been removed.
What is another name for liquid organic manure?

Compost manure

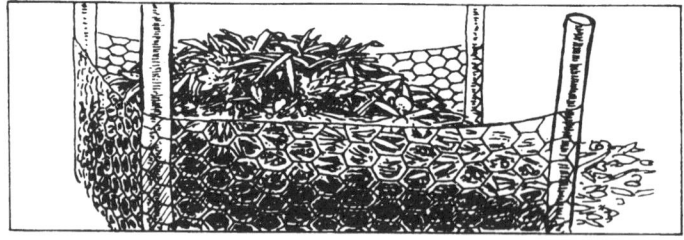

Did you know that compost manure is obtained from plants?
What plant materials are used for composting?
Leaves, soft stems, grass, rejected fruits and vegetables and peelings are all used for composting.
Composting therefore is an easy method of obtaining manure for your home or school garden.

Now study the diagram below and name the materials which are used in composting.

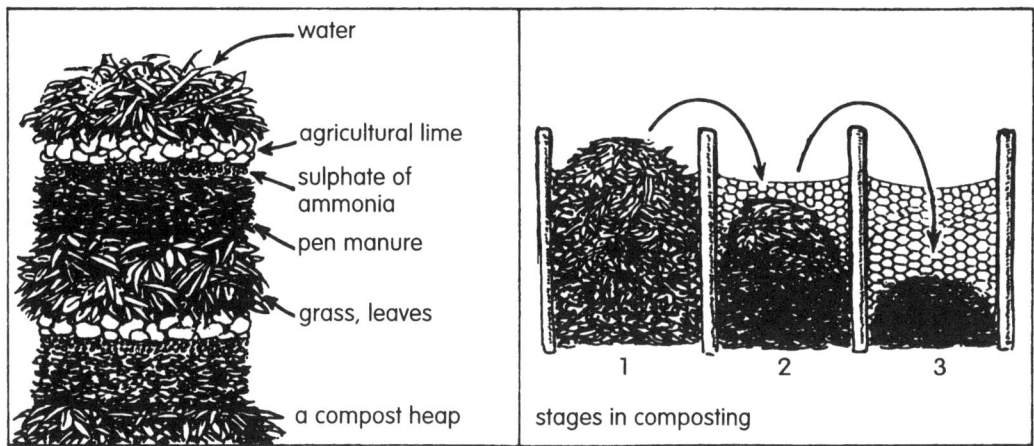

Observe that the materials are stacked in a certain order in different layers.
Observe also the three different stages in composting.
Can you name the function or purpose of each material? Find out from the table below:

Material	Function in the compost heap
grass, leaves, etc.	materials to be decomposed
pen manure	used as a starter to supply the micro-organisms
sulphate of ammonia	supplies nitrogen to enable the micro-organisms to multiply
agricultural lime	to provide a suitable pH for the micro-organisms
water	to provide the necessary moisture and humidity for the micro-organisms

Green manure

Did you know that green manure supplies organic matter and nutrients to the soil?
Observe the materials which are used as green manure.
Notice that they are green crops which are ploughed into the soil, for example, grasses and legumes.
Why is it advantageous to use green legume crops for green manuring?
It is because legume crops add or fix nitrogen in the soil.
Now turn to your Workbook and do **Activity 19**.

(ii) *Inorganic manures*
What are inorganic manures?
How are they obtained?
Inorganic manures are generally referred to as fertilisers.
They are manufactured in fertiliser plants or factories using chemical processes.
Mineral rocks, clay and chemicals are used for making fertilisers or inorganic manures.

Here are some different types of inorganic manures. Study them carefully and try to describe each one.

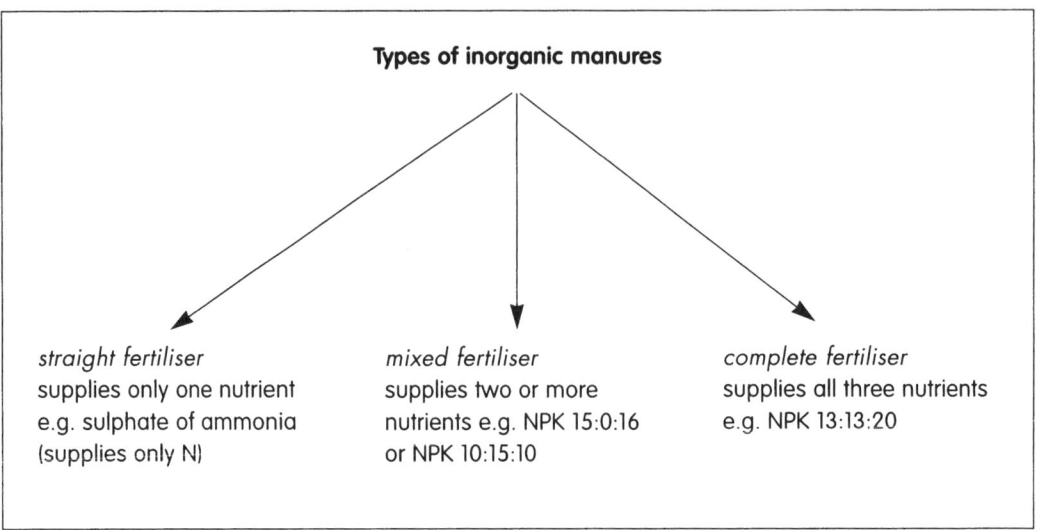

Types of inorganic manures

- *straight fertiliser* supplies only one nutrient e.g. sulphate of ammonia (supplies only N)
- *mixed fertiliser* supplies two or more nutrients e.g. NPK 15:0:16 or NPK 10:15:10
- *complete fertiliser* supplies all three nutrients e.g. NPK 13:13:20

Your teacher will provide you with fertilisers such as sulphate of ammonia, urea, superphosphates, muriate of potash and NPK for you to examine and identify.
In each case, try to find out what nutrients the fertilisers supply and then classify them into the three groups – straight, mixed and complete.

When selecting fertilisers, it is important to choose the type that is most suitable for the particular crop. Carefully study the table below which gives the names of some fertilisers that are most suitable for leafy crops, root crops and fruit crops.

Fertilisers most suitable for leafy crops (e.g. lettuce, patchoi, cabbage)	*Fertilisers best suited for root crops (e.g. yam, cassava, sweet potato)*	*Fertilisers most suitable for fruit crops (e.g. coconuts, tomato, melongene, citrus)*
sulphate of ammonia urea nitrochalk or cal nitro NPK (22:11:11)	single superphosphate triple superphosphate NPK (10:15:10) NPK (10:20:10)	muriate of potash NPK (13:13:20) NPK (12:12:18) NPK (15:0:16)

Now do **Activity 20** in your Workbook.

(d) The nitrogen cycle
Of the three major nutrients, nitrogen is the most important for crop plants. However, because of the heavy rainfall in the Caribbean and tropical countries generally, nitrogen is readily lost from the soil through leaching.
Nitrogen is also lost to the atmosphere and removed from the soil by crop plants.
Therefore, farmers have constantly to add nitrogen-supplying materials (manures) to the soil.

Since nitrogen is so very important in our soils, it is important for you to have an understanding of the *nitrogen cycle*.

A diagram of the nitrogen cycle is given below.
Your teacher will explain to you how it works.

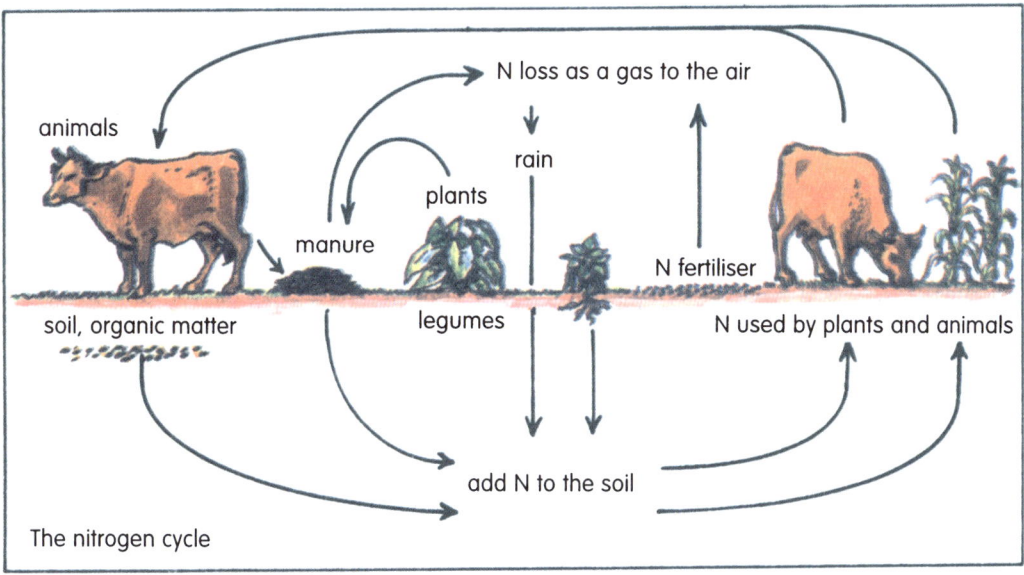

The nitrogen cycle

(e) Storage of manures
(i) Organic
Look at the diagrams below:

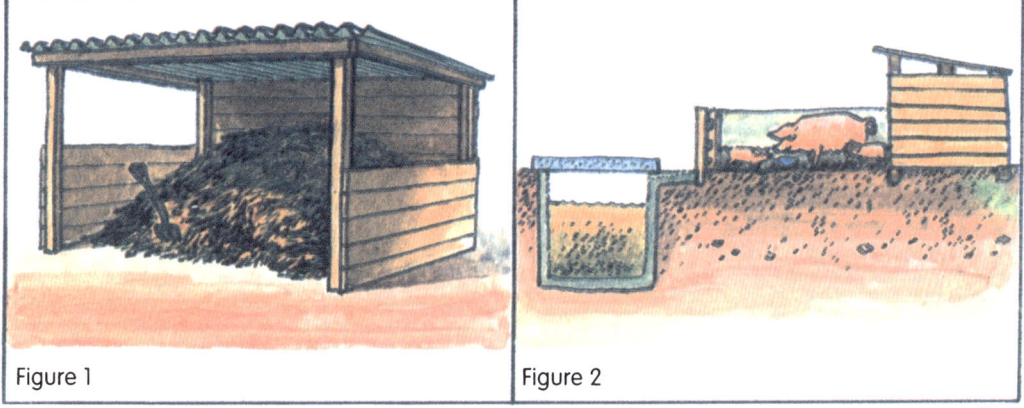

Figure 1

Figure 2

Figure 1 shows how well-rotted pen manure and compost manure should be stored.
Observe the protective wall (brick or board) and the galvanised iron roof on the shed or building.
Can you explain why it is necessary to protect the manure from the elements (sun, rain)?
It is because nutrients can be lost to the atmosphere and by leaching.

Figure 2 shows an underground concrete slurry pit.
Why is the pit concreted and covered?
It is concreted to prevent the nutrients from the liquid manure being lost from the pit.
The covering or lid protects the slurry from the elements, especially the rain which could make the pit overflow, causing loss.

(ii) Inorganic
The diagrams below show you how inorganic manures should be stored:

Observe, firstly, that fertilisers are also protected from the elements. Why is this?
Observe that the bags of fertilisers are stacked on a rack or pallet and notice that these are tied after use. Why? The bags are tied to prevent the fertiliser from absorbing moisture from the air.
Moisture causes fertilisers to become soft. In this state, nutrients are lost to the atmosphere and it becomes difficult for us to apply the moist fertiliser to the soil.
Observe also that fertilisers can be stored in plastic containers (buckets) and polythene bags. Can you explain why tins or metal containers should not be used? It is because fertilisers are corrosive, that is, they can cause tins and iron containers to rust and deteriorate.

(f) Application of manures
We have learnt previously that manures (organic and inorganic) can be applied to the soil at land preparation time and at planting time.
At what other time can manures be applied?
Manures can also be applied during the life of crops.

Now study the chart below. It gives us some key points to bear in mind when applying manures.

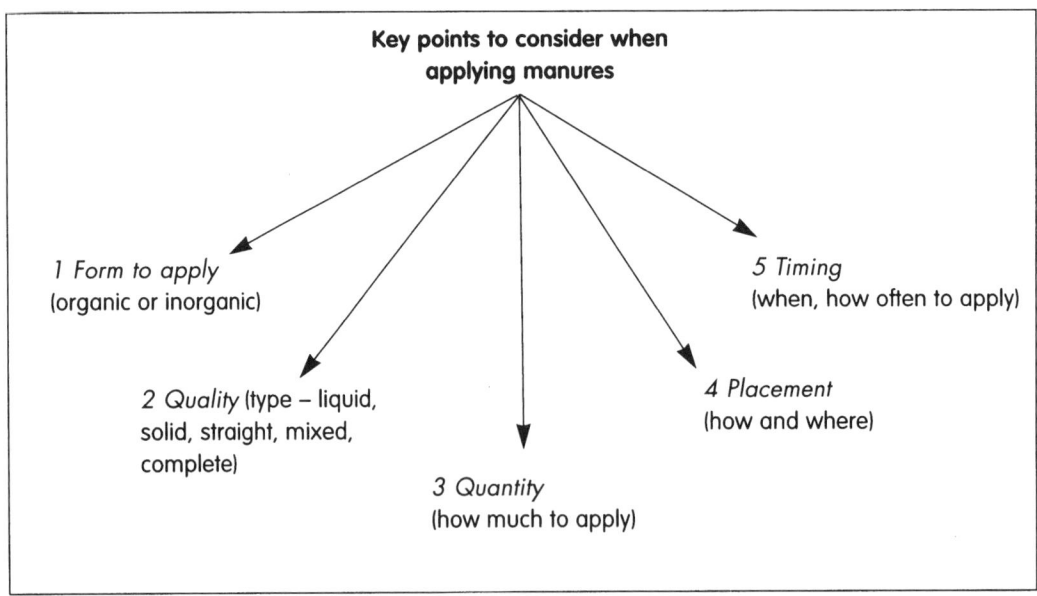

1 Form to apply: decide on whether it is an organic or inorganic manure you wish to use.
2 Quality: choose the type of manure which is most suitable for the crop. You may choose a liquid (organic) manure for pastures, a straight fertiliser such as sulphate of ammonia for growing crops or a complete fertiliser such as NPK (13:13:20) for a bearing tomato crop.
3 Quantity: look carefully at the height or stage of the crop and decide on how much manure to apply; it is advisable to apply small quantities in two or three applications rather than all at one time.
4 Placement: look at the stage of the crop and decide whether the manure should be applied on the soil surface or whether it should be placed or worked into the soil.
Avoid damaging the roots of crops, and place manure where the crop plants can obtain it. (Revise the methods of applying fertilisers: see Book 2, page 34.)
5 Timing: look at the soil type as well as the age and state of the crop. These will help you to decide on how often to apply manures. You may have to apply small amounts every 3–4 weeks, depending on the crop.
Decide on the best time to apply the manures, e.g. just before flowering; after harvesting; just before pruning; when the rainy season is about to start.

Now turn to your Workbook and do **Activity 21**.

5 Weed control
We learnt in Book 2 that weeds are harmful to crop plants in the following ways:
— weeds rob crop plants of nutrients, water, light and space;
— weeds harbour pests and diseases which may later attack crop plants.

Do you remember what a weed is?
A weed is any plant that is unwanted in the place where it happens to be growing.
Now, point out and name some common weeds in your school or home garden.

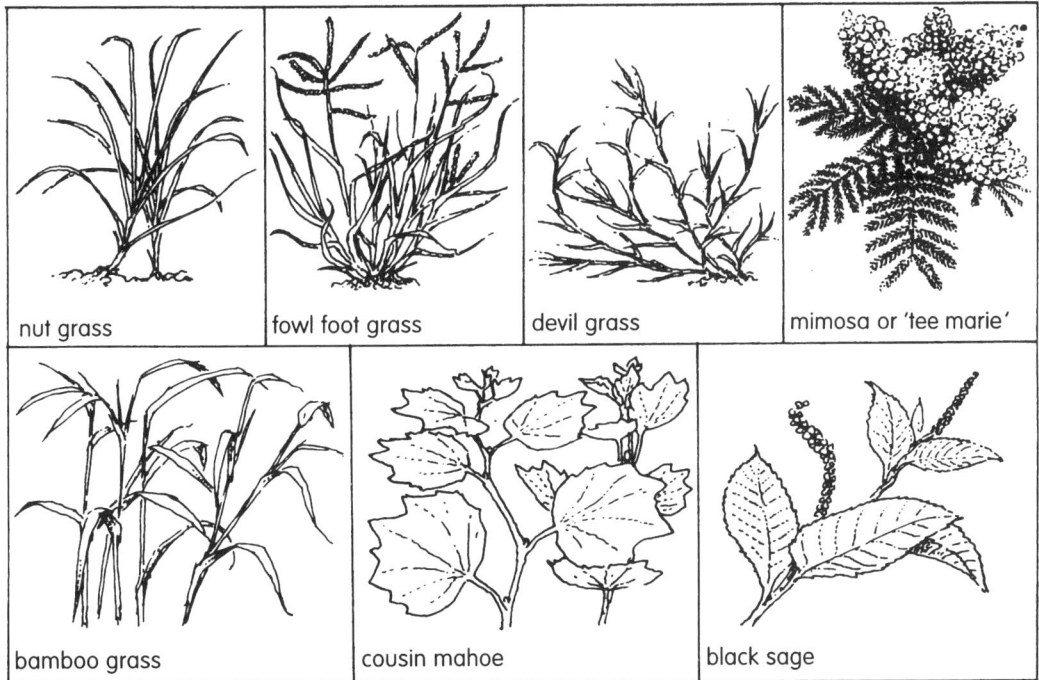

nut grass | fowl foot grass | devil grass | mimosa or 'tee marie'
bamboo grass | cousin mahoe | black sage

Since weeds are so very harmful to crop plants you will agree that we should try to control them.

Here are some common methods we should use to control weeds.
Find out some of the advantages of each method from the chart.

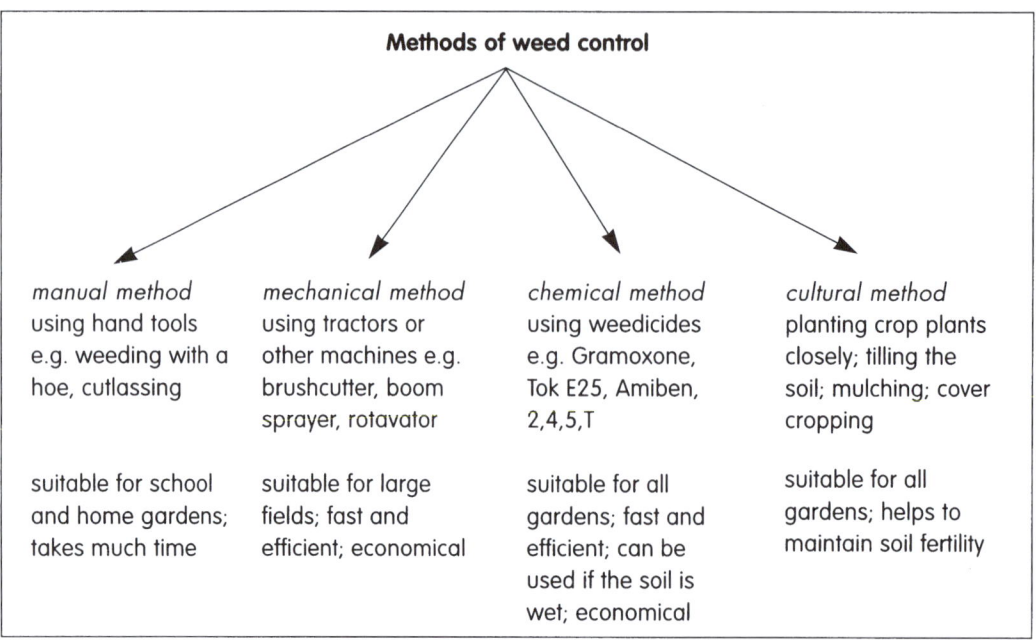

Methods of weed control

manual method
using hand tools
e.g. weeding with a hoe, cutlassing

suitable for school and home gardens; takes much time

mechanical method
using tractors or other machines e.g. brushcutter, boom sprayer, rotavator

suitable for large fields; fast and efficient; economical

chemical method
using weedicides e.g. Gramoxone, Tok E25, Amiben, 2,4,5,T

suitable for all gardens; fast and efficient; can be used if the soil is wet; economical

cultural method
planting crop plants closely; tilling the soil; mulching; cover cropping

suitable for all gardens; helps to maintain soil fertility

Now check the diagrams below to find out the names of tools and equipment which are used for controlling weeds.

hoe | cutlass | knapsack sprayer

tractor | rake | handfork

brushcutter | rotavator | trowel | boom sprayer

Which of these tools and equipment are used for:
(a) manual weed control
(b) mechanical weed control
(c) chemical weed control
(d) cultural weed control?

Now do **Activity 22** in your Workbook.

Unit 4 **Crop production**

1 Factors affecting crop production
Study the chart below. It shows us that successful crop production depends on several factors. Name the factors which are responsible for successful crop production.

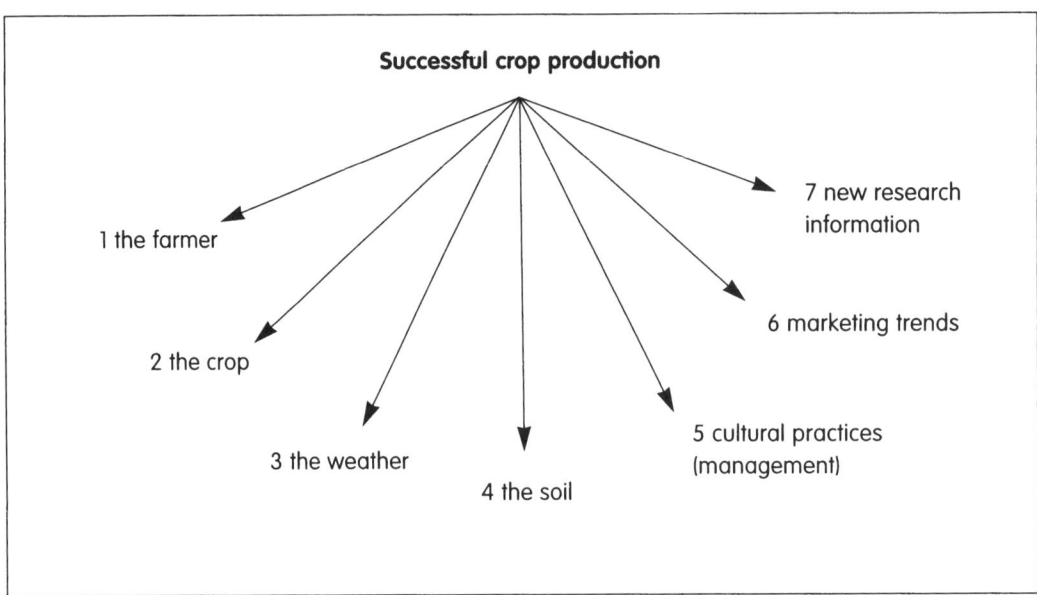

Can you explain briefly how each of these factors affects crop production?
Your teacher will help you if you are not sure.

2 Steps for growing a crop
Now let us look at all the points we should consider when growing a crop.
  1 *Recording*: keep records of all dates and operations.
  2 *Land preparation*: land clearing, ploughing, refining etc.
  3 *Crop varieties*: choice of high yielding, early bearing, disease resistant, or good quality.
  4 *Planting material*: selection and treatment of seeds, seedlings, cuttings, etc.
  5 *Planting distances or spacing*: distances apart in the row and distances between the rows.
  6 *Manuring*: organic or inorganic, type, quantity to apply, when, how.
  7 *Weed control*: methods to use, when, how often.
  8 *Pest and disease control*: methods to use, name of pest, disease and chemical, how much chemical to use, how often.
  9 *Cultural practices*: mulching, drainage, pruning, staking, moulding, which to use and why.
10 *Harvesting*: stage to harvest – snap stage (as in beans), green mature stage, or dry; how to harvest.
11 *Preparation and storage*: trimming, washing, grading, packaging, cold storage.
12 *Marketing*: transport, displaying, selling, wholesale or retail.
13 *Use of produce*: for stewing, currying, pickling etc.
14 *Nutritional value*: main nutrients crop contains – vitamins, proteins, carbohydrates, minerals.

3 Major groups of crops

Given below is a list of the major groups of crops.

For each group you will find that a number of the management practices are similar, so it is easier for you to remember the various steps and management practices.

Notice that only some examples of crops are named in each group. With the help of your teacher, add other examples to the list.

1 Vegetables
(i) Leafy vegetables

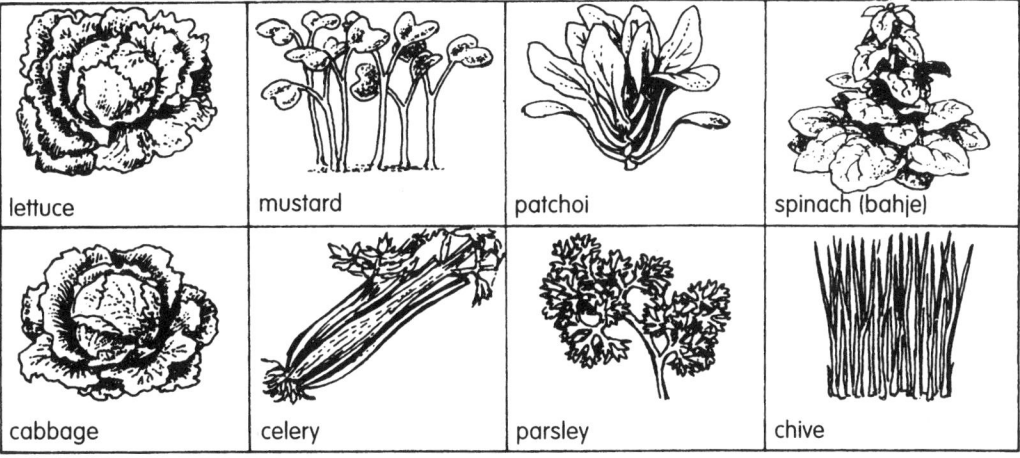

(ii) Root vegetables

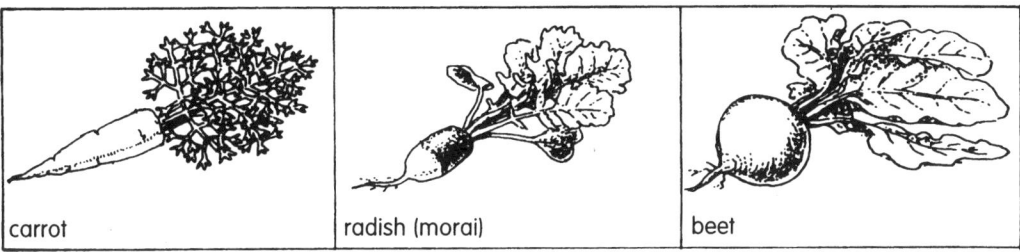

(iii) Fruit vegetables

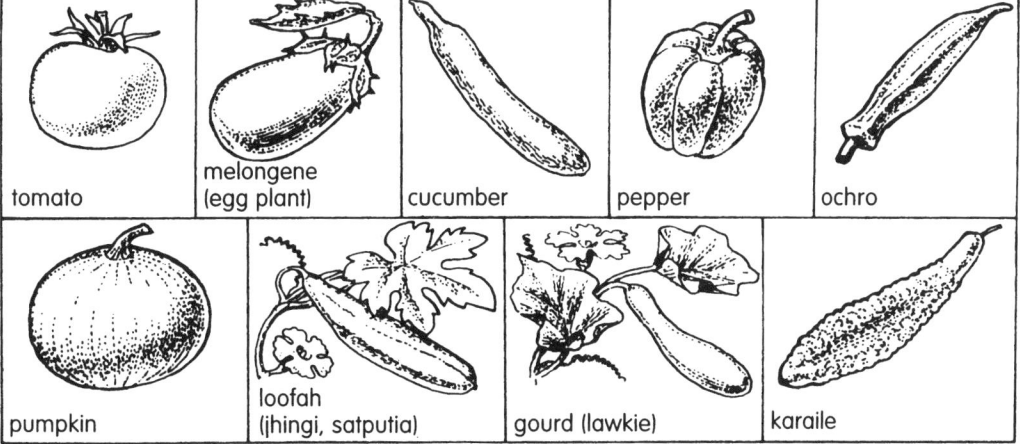

59

2 Root crops

3 Legume crops

4 Cereal crops

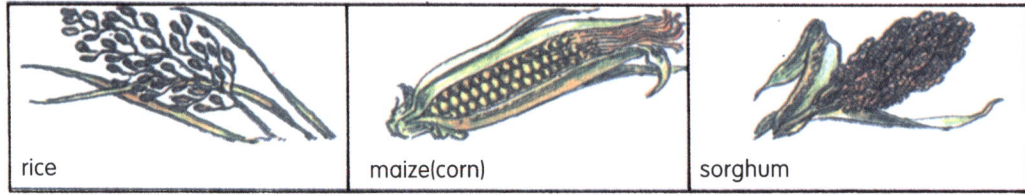

5 Plantation crops

6 Ornamental crops

7 Fodder crops

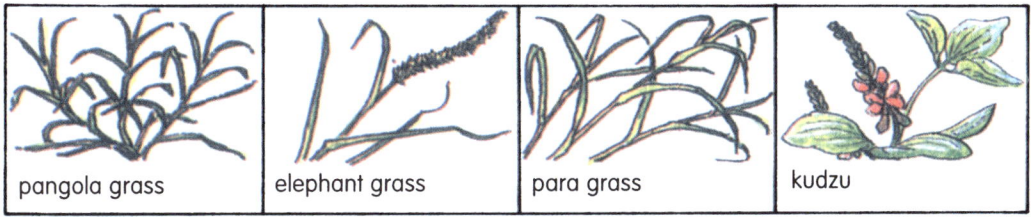

8 Forest crops

Now turn to your Workbook and do **Activity 23**.

4 Cultivation of crops
(a) Lettuce (a leafy vegetable)
Land preparation: Clear the land, plough (fork), rotavate (refine) the soil to a fine tilth.
Dig drains and make raised beds about 1 m wide and 3 m long.
Spread well-rotted pen or compost manure evenly and incorporate it into the soil.
Spray soil with soil insecticide, e.g. Diazinon – 30 ml in 4.5 l water.
Crop varieties: Bronze Mignonette, Iceberg, Butter-crunch, Green Mignonette, Empire, Trinity.
Planting material: Hardened seedlings which are about 3 weeks of age.
Planting distances (spacing): Transplant seedlings 25 cm apart in rows 25 cm apart.
Manuring: Apply 10–15 gm of sulphate of ammonia per plant one week after planting, then again two weeks later. Dissolve 10 gm urea or nutrex in 4.5 l water and spray the leaves once a week.
Weed control: Hand-weeding is suitable; remove weeds as they appear. Light tillage helps to control weeds.
Pest and disease control: Spray with an insecticide, e.g. Malathion for caterpillar attack – use 10–15 ml in 4.5 l water. For fungal attack use a fungicide such as Kocide 101 – 25–30 gms in 4.5 l water.
Cultural practices: Irrigate or water the crop regularly. Lightly dig up or loosen the soil for aeration and to control weeds. Mulch in the dry season.
Harvesting: Lettuce heads are ready for harvesting in 4–6 weeks.
Preparation and storage: Trim off roots and old bottom leaves. Keep in a cool place and spray with cold water. (Do not place in deep freeze.)
Marketing: Grade according to size, package in polythene bags and market immediately.
Use of crop: Used as a salad, eaten raw.
Nutritional value: Mainly vitamins.

Check the Crop Guide on pages 66 and 67 to find out about the cultivation of other leafy vegetables.

Now do **Activity 24** in your Workbook.

(b) Tomato (a fruit vegetable)

Land preparation: Same as for lettuce. (A medium soil tilth can also be used.)

Crop varieties: Floradel, Floralou, Walter, Indian River, Cascade, Kada, Nema, Tropic Boy.

Planting material: Hardened seedlings which are about 3–4 weeks of age.

Planting distances: Transplant seedlings 45 to 60 cm apart in rows 60 to 75 cm apart.

Manuring: Apply pen or compost manure to each hole at time of planting. Apply NPK (13:13:20) one week after planting – use 15–25 gm per plant. At flowering, which is about 4 to 5 weeks after planting, apply NPK (13:13:20), 30–45gm per plant. A third fertiliser application is given 3–4 weeks later. Use NPK (13:13:20) plus sulphate of ammonia (about 30 gm per plant).

Weed control: Uprooting or weeding with a hoe as the weeds appear. Do first weeding about 2 weeks after planting and second weeding 2–3 weeks later.

Pest and disease control: Use soil insecticide such as Diazinon for mole crickets, slugs and cut worms. Use leaf insecticide such as Sevin (20–25 gm in 4.5 l water) for caterpillars.
Use an insecticide such as Monitor for leaf miner attack and a fungicide such as Kocide 101 for fungal attack.

Cultural practices: Stake plants 1–2 weeks after planting. Prune off side shoots, if you wish, and tie stem of plants to stakes. Mould plants and water regularly.

Harvesting: Harvest mature fruits. You can recognise them because the area below the sepals or calyx becomes brown in colour.

Preparation and storage: Place fruits on a large wooden tray in a dry protected area for ripening.

Marketing: Grade the tomatoes and package in polythene bags; they can be sold wholesale or may be retailed for more profit.

Use of crop: For salads, stews, tomato ketchup.

Nutritional value: Mainly vitamins and minerals.

Find out about the cultivation of other fruit vegetables from the Crop Guide.

Now turn to your Workbook and do **Activity 25**.

(c) Carrot (a root vegetable)

Land preparation: Carrots thrive best in a fairly loose, loamy soil. Clear the land, plough (fork), rotavate (refine) the soil to a fine tilth. Make raised beds 3 m long and 1 m wide. Spread about 500 gm NPK (10:15:10) on each bed and incorporate into the soil. (Do not use pen or compost manure for growing carrots, because it causes the roots to be branched or forked.)

Make flat-topped ridges about 20–25 cm wide, 15 cm high and 30 cm apart on the beds, with 3 rows of ridges on each bed.
Crop varieties: Danvers, Chantenay.
Planting material: Seeds.
Planting (sowing the seeds): Make a shallow drill (1 cm deep) along the top centre of each ridge. Mix the seeds with sand in a 1:1 ratio, that is, one part of seeds to one part of sand.
Sow the seeds evenly in the drill and cover over with fine soil to a depth of about 5 mm.
Manuring: About 3 weeks after germination, use NPK (22:11:11) fertiliser and apply it in a band about 4 cm from the plants. Apply about 500 gm per bed. 3 to 4 weeks later apply NPK (10:20:10) in the same way; use about 500 to 700 gm per bed.
Weed control: Use a pre-emergent weedicide such as Tok E25 immediately after planting (about 60 ml in 4.5 l water). After the first true leaves appear, carrot kerosene (stoddard solvent) can be sprayed over the entire bed to control weeds; use about 60 ml carrot kerosene to 4.5 l water.

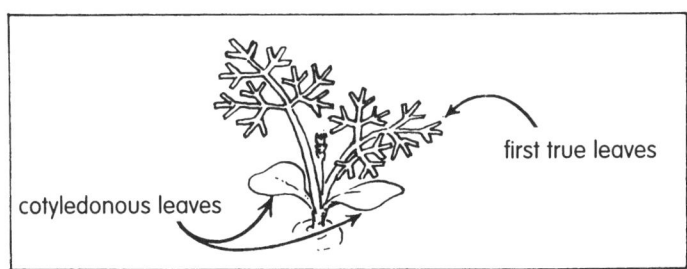

Stop using the chemical at least one month before harvesting to avoid kerosene taste (taints) in the carrots.
Pest and disease control: Caterpillars and other leaf-eating insects can be controlled with Sevin or Malathion. Use a fungicide such as Kocide 101 for fungal attack.
Cultural practices: Thin out seedlings so that they are about 7 cm apart. Water the crop regularly. Lightly loosen the soil to aerate it and to control weeds. Mould or earth-up the plants slightly, and ensure that there is proper drainage.
Harvesting: Carrots are ready for harvesting in 3 to 4 months. At this time the widest diameter of each carrot should be 3–5 cm.
Carrots are normally uprooted.
Soaking the ground just before harvesting helps.
However, a trowel can be used to lift the carrots gently out of the ground.
Preparation and storage: Trim off the leafy tops, wash and air dry, then store in a cool, dry, well-aerated room preferably in crocus bags or baskets. Avoid bruising the carrots.
Marketing: Grade and package the carrots in polythene bags; they can be sold wholesale to groceries and supermarkets or retailed in the open vegetable market.
Use of the crop: For salads and stews. It is eaten raw or boiled and may be sliced, diced or grated.
Nutritional value: Rich in vitamins; also contains minerals.

Check the Crop Guide to find out about the cultivation of other root vegetables.

Now do **Activity 26** in your Workbook.

(d) Yam (a root crop)
Land preparation: Brushcut the land and plough to a depth of about 30 cm. Rotavate the soil to a medium tilth. Broadcast NPK (10:20:10) fertiliser on the soil at the rate of 150–200 gm per square metre (m^2). Also broadcast Furadan granules (30–45 gm per m^2) to control nematodes and other soil pests. Form ridges 30–45 cm high and about 1 m apart.

Crop varieties: White Lisbon, Chinese, Yellow Lisbon, local varieties.
Planting material: Small whole tubers are mostly used for planting. Large yams can be cut into pieces of about 120–150 gm each, but each cut piece must have skin on it otherwise it will not germinate.
Planting distance: Yam tubers are planted 30–45 cm apart on the ridges. Planting should be done in early May.
Manuring: About 2 months after planting, use NPK (10:15:10) and apply 45–60 gm to each plant. Place fertiliser beneath the soil to reduce loss of nutrients. Make 3 or 4 similar applications every 2 months.
Weed control: Immediately after planting, use a pre-emergent weedicide such as Gesaprim (30 gm in 4.5 l water) and spray the soil. Later, weed growth can be controlled by hand weeding or use Gramoxone (60 ml in 4.5 l water) with a shield between the rows.
Pest and disease control: Use an insecticide to control caterpillars and other leaf-eating pests. A fungicide such as Kocide 101 or Tri-Miltox can be used for controlling leaf spots and other fungal diseases. Furadan granules can be used for controlling nematodes.
Cultural practices: Stake yams as shown below. It helps to improve yields.

Harvesting: Yams are harvested 9 to 12 months after planting. Carefully dig out the tubers, wash and air dry.
Preparation and storage: Store tubers in a cool, well-ventilated room. Paste cut or damaged surfaces of tubers with ground limestone or wood-ash before storing.
Marketing: Grade the tubers and sell at different prices.
Use of the crop: Boiled and eaten sliced or mashed; can be processed into 'instant yam'.
Nutritional value: Mainly carbohydrates.

Find out how to cultivate other root crops from your Crop Guide.

Now turn to your Workbook and do **Activity 27**.

(e) Bodi beans (a legume crop)
Land preparation: Same as for lettuce.
Crop varieties: Los Banos Bush Sitao No. 1, local varieties such as Half yard, One yard.
Planting material: Seeds.
Planting distances: For Los Banos plant two seeds about 2 cm deep, 15 cm apart in the row with 50 cm between rows. For climbing varieties, plant seeds 45 cm apart in the row with 75 to 90 cm between rows.
Manuring: Incorporate manure (organic and inorganic) at land preparation time.
Since bodi is a legume crop, it adds nitrogen to the soil. As such it does not normally need nitrogenous fertiliser such as sulphate of ammonia. However, apply the fertiliser to the crop if there is any sign of stunted growth and yellowing.
Weed control: Immediately after planting spray the ground with a pre-emergent weedicide such as Tok E25 (60 ml in 4.5 l water). Hand weeding can be done afterwards or Gramoxone can be applied between the rows using a shield.
Pest and disease control: Same as for carrots.

Cultural practices: Staking has to be done for vining or climbing varieties, but the Los Banos variety does not need stakes. Lightly loosen the soil between the rows for aeration and weed control. Mould or earth-up the plants.

Harvesting: Harvesting is done when the pods are still in the snap stage. Look out for indentation on the pods.

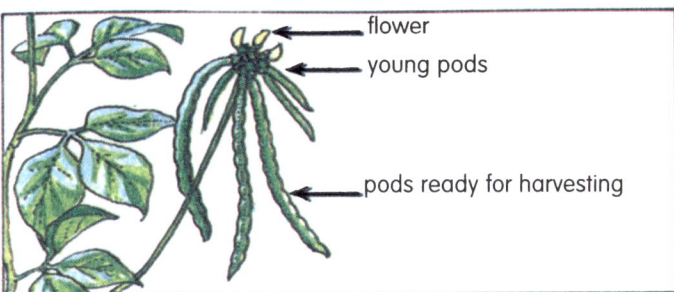

Hold each pod between your thumb and index finger and carefully nip it off from the plant. Avoid damaging the young flowers.

Preparation and storage: Store pods in large round baskets or trays in a cool room.

Marketing: Tie pods in small bundles (0.5 kg) or package in polythene bags for convenience when marketing.

Use of the crop: Can be curried, steamed and stewed by itself or mixed with meat, fish and other vegetables.

Nutritional value: Rich in proteins and vitamins.

Check your Crop Guide to find out about the cultivation of other legume crops.

Now do **Activity 28** in your Workbook.

(f) Maize or corn (a cereal crop)

Land preparation: Clear the land, plough (fork), and rotavate (refine) the soil to a fine or medium tilth. Broadcast NPK (22:11:11) at the rate of 150–200 gm per square metre (m^2).

Beds can be about 3–5 m wide and any length. Camber the beds slightly for drainage.

Apply Diazinon for soil pests (30 ml in 4.5 l water).

Crop varieties: Hybrid corn, such as X306, Pioneer Hybrid; local corn, such as Chaguaramas Corn, St Augustine selection.

Planting material: Seeds.

Planting distances: 25–30 cm in the row and 75 cm between the rows. Plant seeds about 3–5 cm deep.

Manuring: Corn is a very greedy or voracious user of nutrients, especially nitrogen. About 2 weeks after germination apply sulphate of ammonia, or NPK (22:11:11) in a band 10 cm away from the plants. Use 30–40 gm of fertiliser along a distance of one metre (1 m).

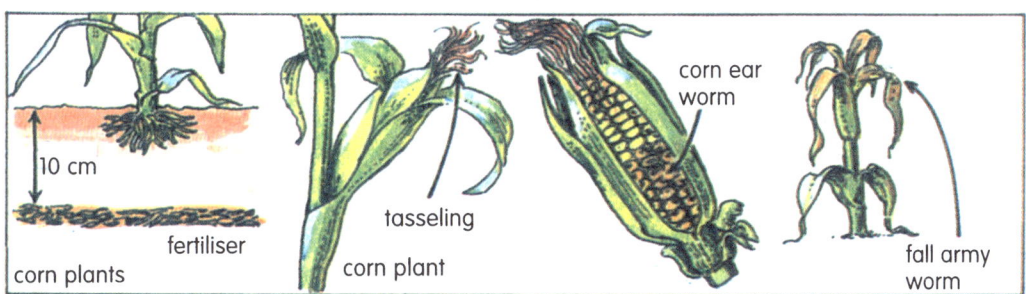

(Continued on page 68)

Crop Guide

No.	Crop	Group crop	Common varieties	Planting material used
1	patchoi	leafy vegetable	Spoon leaf, White Petiole, Green Petiole	seedlings
2	spinach	leafy vegetable	Tall red stem, Tall white stem	seedlings
3	cabbage	leafy vegetable	Tropicana, Caribbean Queen, Fortuna	seedlings
4	cauliflower	flower vegetable	Early Market, Early Patna, Kono 45	seedlings
5	celery	leafy vegetable	Dewcrisp, Tall Utah, K Strain	seedlings
6	chive	leafy vegetable	Japanese bunching, local	seedlings
7	onion	leafy vegetable	Texas Early Grano, Granex F1	seedlings
8	melongene	fruit vegetable	Long Purple, Oval Purple	seedlings
9	sweet pepper	fruit vegetable	California Wonder, King Henry	seedlings
10	hot pepper	fruit vegetable	Scotch Bonnet, Caribbean Red, Habanero	seedlings
11	ochro	fruit vegetable	Six weeks, Clemson spineless, Perkins dwarf	seeds
12	sorrel	fruit vegetable	Red dwarf, White dwarf	seeds
13	cucumber	fruit vegetable	Ashley, Straight eight, Poinsett, Quest	seeds
14	pumpkin	fruit vegetable	Big Max, Crapaud Back (local), Jack O' Lantern	seeds
15	melon	fruit vegetable	Royal Sweet, Regency, Lady Di	seeds
16	radish	root vegetable	Horse radish (red); white radish (morai)	seeds
17	beet	root vegetable	Detroit Dark Red, Ruby Queen	seeds
18	cassava	root crop	Butter Stick, Red Stick, M Col., MX	stem cuttings
19	sweet potato	root crop	049, local	stem cuttings
20	eddo	root crop	local	corms
21	dasheen	root crop	local	corms
22	tannia	root crop	local	corms
23	salad beans	legume	Contender, Red Kidney bean, Harvester	seeds
24	bodi beans (climber)	legume	Green Arrows, One yard, Half yard (local)	seeds
25	bodi beans (dwarf)	legume	Los Banos Bush Sitao No. 1	seeds
26	pigeon peas (tall)	legume	Tobago Peas, Semi-dwarf	seeds
27	pigeon peas (dwarf)	legume	Dwarf	seeds
28	rice	cereal	Dima, IR8, Sughandi	seedlings
29	sorghum	cereal	Pioneer hybrid 820, Dorado M	seeds
30	sugar cane	plantation crop	B41227, HJ5741	stem cuttings
31	cocoa	plantation crop	Criollo, Forastero	seedlings; rooted stem cuttings
32	coffee	plantation crop	Arabica, Robusta	seedlings
33	coconuts	plantation crop	Tall type, Malayan dwarf (Chinese)	seedlings
34	citrus (grapefruit)	plantation crop	White Marsh, Pink Marsh	budded plant
35	oranges	plantation crop	Valencia, Parson Brown, Navel	budded plant
36	limes	plantation crop	West Indian, Tahiti	budded plant

Planting distances in rows	Planting distances between rows	Depth of planting seeds	Season or best time to plant	Ready for harvest
30 cm	30 cm	5–10 mm	all year round	7–8 weeks
30–45 cm	30–45 cm	5–10 mm	all year round	6–12 weeks
30–45 cm	45–60 cm	5–10 mm	Dec. to May	3–3½ months
45–60 cm	60–75 cm	5–10 mm	Dec. to May	3–4 months
25–30 cm	25–30 cm	3–5 mm	all year round	8–12 weeks
20 cm	20 cm	3–5 mm	May to Dec.	3–4 months
7–10 cm	25–30 cm	5–10 mm	Oct. to Dec.	3–4 months
100–120 cm	120 cm	5–10 mm	all year round	3–6 months
35–45 cm	45–60 cm	5–10 mm	all year round	2½–3 months
100–120 cm	120 cm	5–10 mm	all year round	3–6 months
60 cm	75–90 cm	1–2 cm	all year round	1¾–4 months
120 cm	150 cm	1–2 cm	May to June	6–10 months
90 cm	120–150 cm	2–3 cm	Sept. to Jan.	2–3 months
2 m	3 m	2–3 cm	Sept. to Jan.	3½–4 months
90 cm	120–150 cm	2–3 cm	Dec. to Feb.	3–3½ months
7 cm	30 cm	5–10 mm	all year round	2½–3 months
7 cm	25 cm	5–10 mm	all year round	2½–3 months
75–100 cm	120 cm	—	May to June	6–12 months
35–40 cm	75 cm	—	Sept. to Nov.	4–5 months
60 cm	75–90 cm	—	May to June	6–9 months
60 cm	90 cm	—	May to June	6–9 months
60 cm	90 cm	—	May to June	6–9 months
10 cm	45–50 cm	3–5 cm	Oct. to Feb.	8–12 weeks
60 cm	90 cm	3–4 cm	all year round	10–14 weeks
15 cm	45–50 cm	3–4 cm	all year round	10–14 weeks
120–150 cm	120–150 cm	2–3 cm	May to June	8–9 months
60 cm	75 cm	2–3 cm	Nov. to Jan.	4–5 months
25–30 cm	25–30 cm	—	June to July	5–6 months
30 cm	75 cm	2–3 cm	Oct. to Dec.	3–3½ months
45 cm	90 cm	—	May to June	8–10 months
3–4 m	3–4 m	5–7 cm	June to July	3–4 years
2–3 m	3 m	3–5 cm	June to July	3–4 years
5 m	7–8 m	15 cm	June to July	3–3½ years
8 m	8 m	—	June to July	3–3½ years
6 m	6 m	—	June to July	3–3½ years
5 m	5 m	—	June to July	3–3½ years

Three weeks later give a similar application. Just before tasseling or flowering starts apply NPK (12:12:18 or 13:13:20) at the rate of 40–50 gm per metre.

Weed control: Similar to that for yams.

Pest and disease control: Fall army worms and corn ear worms (caterpillars) can be controlled using insecticides such as Sevin and Malathion. Control soil pests with Diazinon.

For leaf spots and other fungal attack use fungicides such as Kocide 101 or Tri-Miltox.

Cultural practices: Irrigate or water regularly. Lightly loosen the soil between rows for aeration and weed control.

Mould or earth-up plants and ensure that there is proper drainage.

Harvesting: Depending on variety, corn is ready for harvesting in the green stage in 10 to 12 weeks. Carefully slit sheaths on one side of a few cobs and examine to find out whether the green corn is ready for harvesting. When harvesting, simply break off the cobs from the corn plants.

Preparation and storage: Remove sheaths from corn cobs and trim the ends. Place in a cool room.

Marketing: Grade and package corn cobs for marketing.

Use of crop: Cobs are normally eaten boiled or roasted. Green corn can also be used in soups. Dry corn is grated and made into corn flour. Corn oil is also extracted from the dry grains. Dry corn as well as corn stalks are used for feeding livestock.

Nutritional value: Mainly protein but also contains carbohydrates, fats, vitamins and minerals.

Find out about the cultivation of other cereal crops from your Crop Guide.

Now turn to your Workbook and do **Activity 29**.

(g) Bananas (a plantation crop)

Land preparation: Similar to that for maize (corn): make beds about 4 m wide and broadcast Furadan to control nematodes and other soil pests.

Crop varieties: Dwarf Cavendish, Gros Michel, Lacatan, Robusta, Valerie, Plantain, local varieties such as silk fig, moko, sucrier.

Planting materials: Suckers, bull-heads.

Planting distances: Pare or clean suckers and dip (treat) them in a soil insecticide solution, such as Diazinon. Plant suckers 2 m apart in the row with about 3 m between rows.

Manuring: About 1 month after planting apply NPK (22:11:11) at the rate of 250 gm per plant.

Use NPK (22:11:11) and NPK (12:12:18) alternately every 2 to 3 months. Start with 250 gm per plant and gradually increase to 500–600 gm per plant or stool. Bananas need lots of potash. Pen manure and compost manure can also be applied around the stools.

Weed control: Gramoxone with a shield can be used to control weeds between the rows. Brushcutting can also be done using a mower or brushing cutlass.

Pest and disease control: Same as for yams.

Cultural practices: Prune suckers regularly (every 4–6 weeks) so that each stool or mat contains one 'mother' plant and one follower. Pruning should result in larger bunches of bananas (high yields). Ensure that there is proper field drainage. Remove flower cushions after a bunch is formed and 'sleeve' with polythene bags to protect bunches from injury.

flower cushion

bunch 'sleeved'

Harvesting: Harvest mature bunches. Avoid damaging or bruising the fruits.

Preparation and storage: De-hand bunches carefully. Wash off latex or 'milk' and then dip in a fungicide solution to prevent stem rot.

Air dry fruits, pack in cardboard boxes and store in a dry warm area for ripening.

Marketing: The green fruits are packed in cardboard boxes and exported.

Locally, fruits are sold by the bunches or by the 'hand', both in the green and ripened state.

Use of the crop: Eaten as a ripe fruit. Green fruits are boiled or cooked in a number of ways and commonly used in fish broth. Discarded or rejected fruits can be fed to pigs.

Nutritional value: Rich in minerals such as iron and calcium; also contains carbohydrates and vitamins.

Check your Crop Guide to find out about the cultivation of other plantation crops.

Now do **Activity 30** in your Workbook.

Unit 5 **Livestock rearing**

1 Factors affecting livestock rearing
Carefully study the chart below:

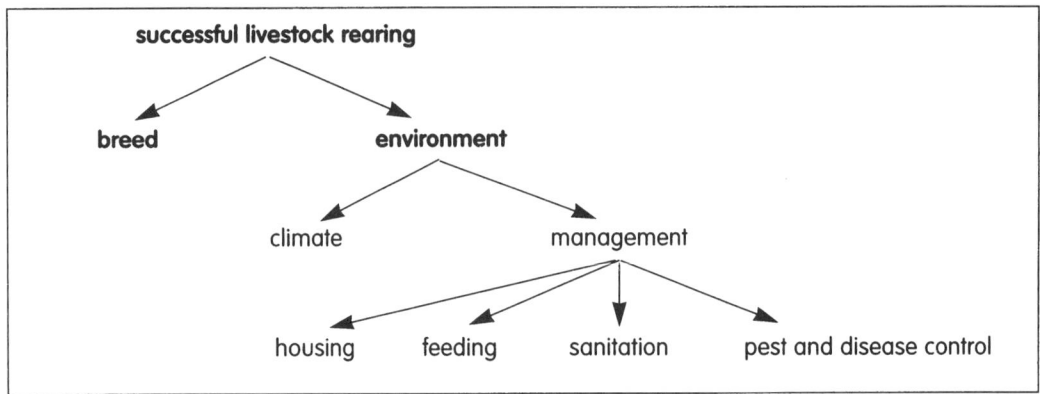

Observe that successful livestock rearing depends on two major factors. Name them.

For successful livestock production, we must choose suitable *breeds* of animals and provide the right kind of *environment* for them.

Can you explain what is meant by 'environment'?
Notice that the *environment* refers to
(a) the *natural* conditions under which livestock are reared, for example *climate*; and
(b) the *artificial* or man-made conditions, that is, the *management* practices which we normally use for rearing livestock.

In order to provide the right environment for our animals, we must
(a) house them properly
(b) feed and water them regularly
(c) always keep the animals and their pens in a sanitary condition, and
(d) protect our animals from pests and diseases.

2 Breeds and breeding
(a) Breeds
Visit a poultry farm with your teacher to look at different breeds of poultry, such as Rhode Island Reds and White Leghorns.

Rhode Island Reds | White Leghorns

Carefully observe and describe the features of a White Leghorn.
— What colour is it?
— Observe its body size. What is its weight?
— What colour of eggs does it lay?
— Look at other White Leghorns. Are they similar?

Now observe and describe the features of the Rhode Island Reds in a similar way.
Are the features the same as those of the Leghorns?

What, would you say, are White Leghorns and Rhode Island Reds?
Both White Leghorns and Rhode Island Reds are different *breeds* of fowls which are reared for eggs.
We have noticed that all White Leghorns are white in colour, small in body size, weigh about 1–1½ kg, and lay white-shelled eggs.
Similarly, we have found out that all Rhode Island Reds are brownish-red in colour, large in body size, weigh about 1½–2 kg and lay brown-shelled eggs.

Can you now explain what is meant by the word 'breed'?
The word 'breed' is a term used to describe a group of animals which have similar features.

It is important for us to know, however, that some breeds of animals may be better than other breeds for some purposes.
Can you explain in what ways breeds are different?
Apart from their differences in colour and appearance, here are some other ways in which breeds are different:
(i) Some breeds have a larger body size than other breeds, e.g. Rhode Island Reds are larger than White Leghorns.
(ii) Some breeds grow faster and mature earlier than other breeds, e.g. Vantress Cross Broilers as compared to Local or Common fowls; Pekin Ducks grow faster and mature earlier than Local or Common ducks.
(iii) Some breeds produce more than other breeds, e.g. Holstein cattle produce more milk than Zebu cattle.
(iv) Some breeds are more resistant to pests and diseases, e.g. Zebu cattle as compared to Holstein cattle.
(v) Some breeds can withstand high temperatures or tropical heat better than other breeds, e.g. Zebu cattle as compared to Holstein cattle.

Now, discuss with your class why it is important for us to choose the best breeds of animals for rearing.

Now turn to your Workbook and do **Activity 31**.

(b) Breeding
(i) What is breeding?
Breeding is the mating of physically mature male and female animals.
We have already learnt that every breed of animal possesses certain specific features. These features are known as *hereditary* characteristics and are passed on from parents to their offspring.

When we breed animals, our main objective is to transfer hereditary characteristics from the two parents (male and female) to the offspring or young. For example, a Zebu bull, which is resistant to pests and diseases, when mated with a high milk-producing Holstein cow gives us an offspring which is resistant to pests and diseases and also a high milk producer.

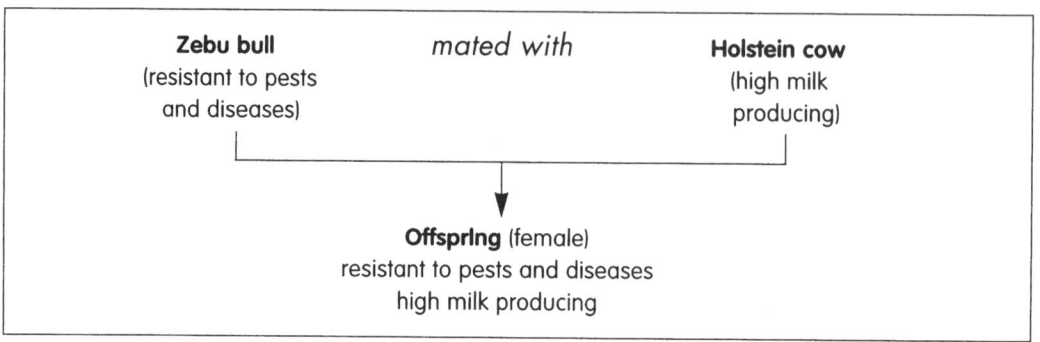

(ii) Types of breeding
(1) Purebreeding
Do you know what is meant by 'purebreeding'?
Study the chart below to find out.

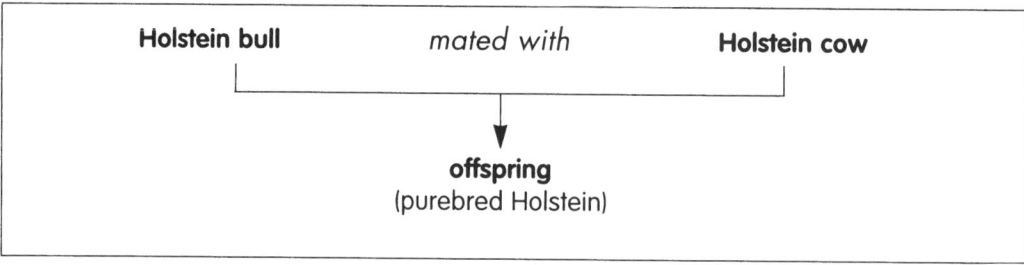

From the chart, you can see that 'purebreeding' is the mating of animals of the *same* breed. The offspring is known as a *purebred*.

(2) Crossbreeding
Now, find out what 'crossbreeding' is from the chart.

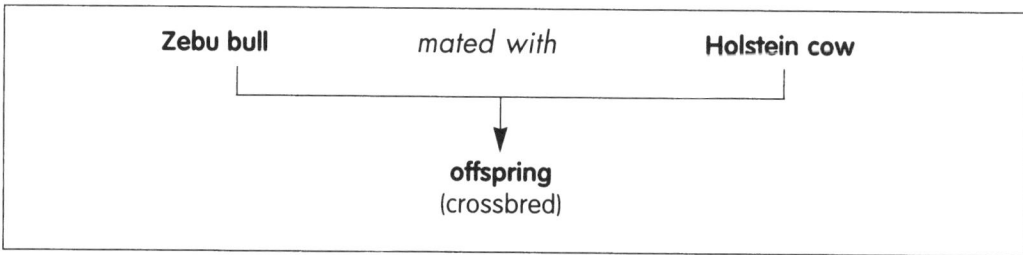

The chart shows that 'crossbreeding' is the mating of animals of *different* breeds.
The offspring is known as a *crossbred*.
With guidance from your teacher draw charts to show other examples of purebreeding and crossbreeding.

(c) The importance of breeding
From what we have studied so far, we can see that breeding is necessary for:
(i) obtaining high producing animals;
(ii) producing animals which are resistant to pests and diseases;
(iii) producing animals which can tolerate or withstand high temperatures or tropical heat;
(iv) obtaining fast growing and early maturing animals.
When breeding, we must produce animals which are suitable for the environment.
Since we live in a *tropical* climate, it is necessary for us to produce or choose breeds of animals which can tolerate tropical conditions: high temperatures, rainfall and humidity.

(d) Breeds of animals

In Book 3, we learnt that animals are reared primarily for meat, milk and eggs. Name some animals which are reared for meat, milk and eggs.

Now study the table below to find out the names of some of the best breeds of animals reared for meat, milk and eggs.

Name of animals	Purpose for which reared	Best breeds for the purpose
Dairy cattle	milk	Holstein, Jamaica Hope; Guernsey, Jersey, Australian Milking Zebu, Brown Swiss
Dairy goats	milk	Saanen, British Alpine, Anglo Nubian, Toggenburg
Beef cattle	meat (beef)	Aberdeen Angus, Charolais, Short Horn, Buffalypso, Jamaica Red, Charbray
Pigs	meat (pork)	Landrace, Large White, Hampshire, Duroc
Goats	meat	Saanen, British Alpine, Anglo Nubian, Toggenburg
Sheep	meat (mutton)	Barbados Black Belly, Nellore, Black Head Persian
Fowls (broilers)	meat (chicken)	Rhode Island, Plymouth Rock, New Hampshire, Vantress Cross, Hyline (strains of broilers)
Ducks	meat (duck)	Pekin, Rouen, Muscovy, Mallard, local
Geese	meat (geese)	Emden, Snow Goose, African
Turkeys	meat (turkey)	Bourbon Red, Black, Slate, Jersey Buff
Rabbits	meat (rabbit)	Flemish Giant, New Zealand White, Silver Fox, Palamino, Belgian Hares, Angora
Fish	meat (fish)	Tilapia, Cascadura
Fowls (layers)	eggs	White Leghorns, Rhode Island Red, Plymouth Rock, Starcross Shavers, Hyline, Golden Comets, Bovan
Ducks	eggs	Khaki-Campbell, Indian Runner

Now do **Activity 32** in your Workbook.

3 Care of young farm animals

In nature, most female animals take care of their young until they are old enough to take care of themselves.

In farming, however, farmers need to take care of young farm animals.
Let us therefore study how to take care of different kinds of young farm animals.

(a) Chicks

The pictures below show us two different methods of brooding chicks (that is, taking care of chicks during their early life).

Figure 1   Figure 2

Figure 1 shows us that chicks can be brooded *naturally* with a *mother hen.*
Figure 2 shows how chicks are taken care of *artificially*.

In natural brooding, the mother hen takes care of her chicks for about 4 or 5 weeks.
Here are the steps we should take:
1 Keep the hen and her chicks in a poultry house to protect them from the elements (sun, rain) as well as from rats, mongooses and other predators. Observe that the hen provides *warmth* for the chicks as they get under her wing and body feathers.
2 Supply feed and water for the hen and her chicks at all times, so that they may eat and drink whenever they feel like doing so.
3 Wash the feeders and waterers every day to keep them in a sanitary condition. Some different types are shown below:

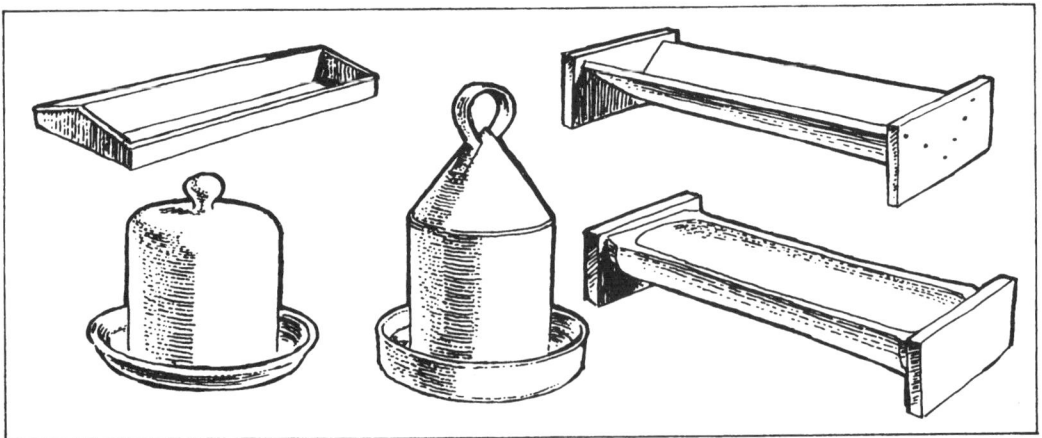

4 Spread bagasse, sawdust, wood-shavings or dry grass on the floor of the poultry pen.
  Do you know why?
  This litter material will absorb the droppings of the birds and so prevent us from having to clean the poultry house every day.
5 Stir or turn over the litter material about twice a week to mix the droppings into it.
6 Vaccinate the chicks with Newcastle/infectious bronchitis vaccine and fowl pox vaccine to protect them from Newcastle disease, infectious bronchitis disease and chicken or fowl pox disease.

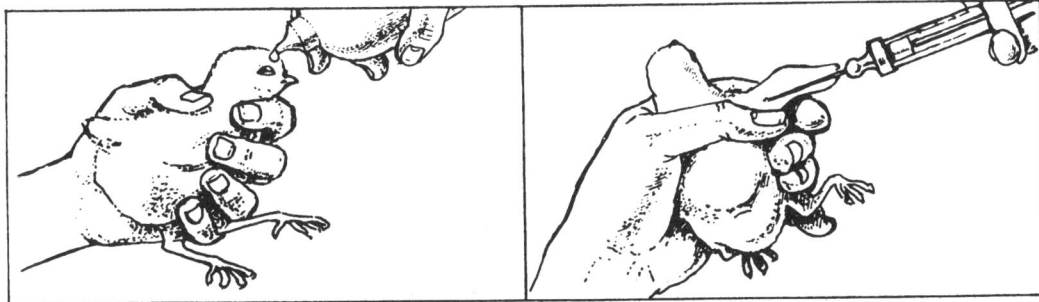

7 Mix the medicines named below in the drinking water to protect the chicks from other diseases and to enable them to grow rapidly and healthily:
   (i) antibiotics and vitamins (e.g. Vita Sol);
   (ii) sulphur drugs (e.g. 5-Sulfas).

Now observe that, in artificial brooding of chicks, there is no mother hen.
How, then, are the young chicks kept warm? They are kept warm by heat from kerosene lamps or electric bulbs (infra-red).

Here are some guidelines for us to follow:
1 During the first week, adjust the temperature of the brooding area or brooder to 35°C.
2 Reduce the temperature of the brooder by 3°C each week until it is about 24°C. Can you explain how to take the temperature of the brooder? Check the diagram below.

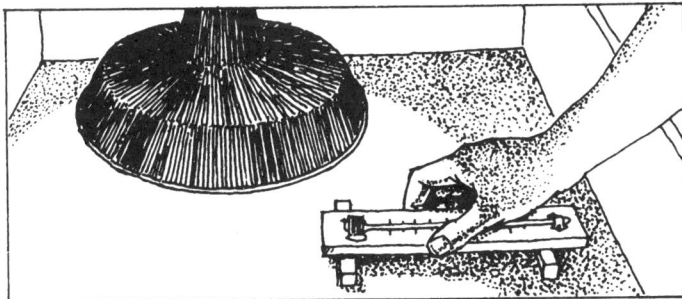

Observe that a thermometer is held about 5 cm above the floor for approximately 3 minutes under the lighted bulb or lamp. The temperature reading is then taken.
How can you adjust the temperature if it is too high or too low?
The temperature is adjusted simply by raising or lowering the bulb or lamp.

3 All the other steps or management practices such as housing and protection, feeding and watering, sanitation and vaccination, are the same as for natural brooding.

Now turn to your Workbook and do **Activity 33**.

(b) Rabbits

Do you remember what the young of a rabbit is called? It is called a kitten.
Have you seen kittens (young rabbits) when they are born? Describe them.
When kittens or young rabbits are born, their bodies are naked, that is, they have no fur. Their naked bodies are reddish in appearance.

Can young rabbits see when they are born or are they blind?
Young rabbits or kittens are blind at birth because they are born with their eyelids closed.
As you can see, young rabbits are quite helpless at birth and require much care and attention.

How will you go about taking care of young rabbits?
Follow these guidelines:
1 Place the pregnant doe or female rabbit in a clean hutch by herself.
2 Make sure that she is protected from the elements and predators.
3 Now, place a nest box in the hutch.

What is the purpose of a nest box?
A nest box provides a place for the doe rabbit to give birth to her young.
Observe that the nest box is large enough to hold or accommodate both the doe rabbit and her kittens.

Observe a doe rabbit which is about to kindle, that is, to give birth to her young.
You will notice that she pulls out fur from her body and places it in the nest box. Why?
The doe rabbit uses her fur to provide a soft, warm bedding for her young.
Count the number of kittens daily to find out whether any are missing.
4 Provide the doe rabbit with fresh, clean feed and water at all times. This is to ensure that she is kept in good health to provide milk for suckling her young. Later, additional feed and water will have to be provided for the young rabbits.
5 Collect fresh herbage (rabbit grass) daily and spread it out on a herbage rack.

Allow the herbage to wilt for one day before feeding. Why?
Wilting reduces the moisture content of the herbage. This helps to prevent watery faeces or scouring in rabbits.

6 Clean the feeder, waterer, hutch and the surroundings every day to keep them in a sanitary condition.
7 Be careful when handling rabbits. The diagram below shows the right way of holding them.

Now do **Activity 34** in your Workbook.

(c) Goats

What is the name of the young of a goat?
How many young does a nanny-goat or she-goat normally produce at birth?
Are the kids as helpless as young rabbits when they are born?

Here are two young kids. How will you go about taking care of them?
Follow these steps:

1 Keep the kids in a clean dry place. Provide a house for them and a place for them to exercise.

2 Make sure that the kids suck the first milk from their mother when they are born. The first milk is known as *colostrum*. It contains substances or antibodies which make the kids more resistant to diseases.

3 Allow the kids to suck milk from their mother or teach them to drink milk from a bucket.
4 Provide clean, fresh water in the pen for the kids.
5 When the kids begin nibbling at grass, provide them with soft herbage. Wilt herbage before feeding. Why?
6 Also provide the kids with a mineral or salt lick.
7 Provide concentrates for the kids for rapid growth.

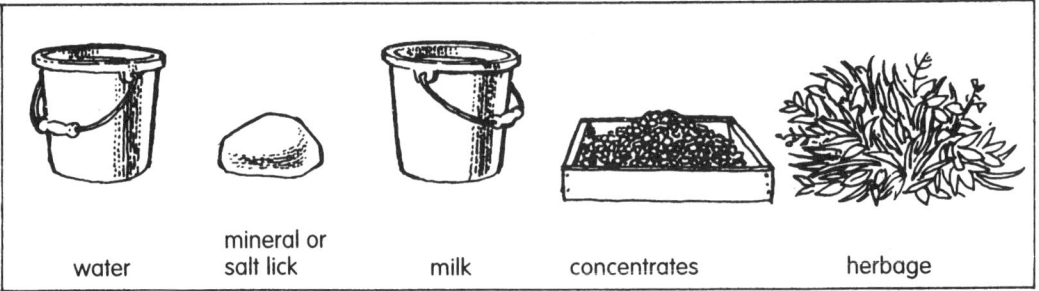

water — mineral or salt lick — milk — concentrates — herbage

8 Clean the pen, feed box and bucket and the surroundings every day.
9 Keep the kids clean. Groom them with a brush and examine them daily for injuries and pests.
10 Allow the kids to exercise every day.

Now visit a farm where calves, lambs and piglets are reared.
Observe carefully and find out how to take care of these young farm animals.

Now turn to your Workbook and do **Activity 35**.

4 Management of growing and adult livestock
Refer to the chart on the first page of this unit.
Observe that management includes four major practices, namely:
1 Housing   2 Feeding   3 Sanitation   4 Pest and disease control
These four practices are important when dealing with any group of farm animals.

(a) Management of broilers
(i) Housing

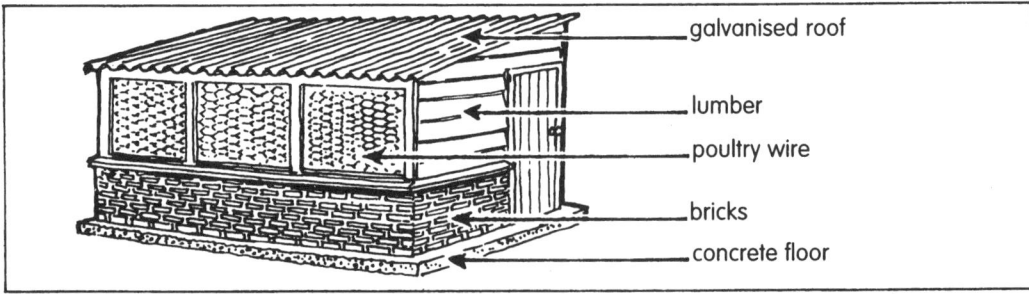

Poultry houses are constructed to protect the growing broiler chicks from the elements and predators.
Look carefully at a poultry house and describe some of its features.
Broiler houses must also be:
— spacious enough to accommodate the birds.
   Each broiler chick requires a floor space of 30 centimetres square $(30 \text{ cm})^2$ or 900 square centimetres (900 cm^2).

Calculate the floor space required for 100 broilers.

How many broilers can be reared in a poultry house 4 m long and 3 m wide?

— well-ventilated to get rid of foul gases and for proper circulation of fresh air.

— well-lighted.

(ii) Feeding

Observe the kinds of feeders and waterers which are used for broilers.

Place a sufficient number of feeders and waterers in the broiler house.

In the brooder, broiler chicks were fed *broiler starter*.

Continue to feed them *broiler starter* in the growing house until they are about 6 weeks old.

If available, *chick grower* can be used during the 4th, 5th and 6th weeks.

Change to *broiler finisher* during the 7th week.

Continue to feed them the *broiler finisher* until they are marketed.

Broilers are normally marketed or sold when they are 8 to 9 weeks old.

Can you say what is the average weight of broiler chicks when they are 8–9 weeks of age?

On average they weigh 2 kg.

In agricultural terms, broiler chicks are fed *ad lib*, that is, they eat and drink whenever they feel like doing so. Feed and water must always be present in the feeders and waterers.

(iii) Pest and disease control

Sanitation

Good sanitation helps to prevent pests and diseases and creates pleasant and healthy surroundings or environment.

Thoroughly clean and disinfect the poultry house before bringing in the chicks.

Clean and disinfect the area around the broiler house regularly.

Wash the waterers daily. Some types of feeders may need daily cleaning as well.

Stir the litter material on the floor about twice a week.

Bury dead chicks and rats or burn them in an incinerator.

Thoroughly clean and disinfect the broiler house after the chicks have been removed.

Medication or medicines

We have already learnt about medicines and vaccines which are given to young broiler chicks (see page 75).

In the growing house, continue to use the medicines below in the drinking water of the birds:

— antibiotics and vitamins (one or two times per week);

— sulphur drugs (two or three times per week).

The antibiotics are used primarily to prevent stress and to control common virus infections.

Sulphur drugs prevent and control internal bleeding in the intestines of the birds.

This disease is commonly known as *coccidiosis* and is caused by tiny organisms called *coccidia*.

Stop using all medicines about 2 weeks before the birds are killed or slaughtered.

(iv) Record keeping
To avoid uncertainty or guesswork, keep proper records of all dates, number of chicks, feed purchased, medicines, money spent, feed consumed weekly, weekly weights of the broilers, profit or loss made and so on.
Remember, records supply us with valuable data or information.

Now do **Activity 36** in your Workbook.

(b) Management of layers
(i) Housing
Layer houses are similar to broiler houses. In addition, layer houses must have these features:
— enough room to allow each layer a floor space of 75 centimetres square – $(75 \text{ cm})^2$;
— nest boxes for the birds to lay (you will observe that layers prefer a slightly darkened area to lay);
— a roost for the birds to sleep at night.

(ii) Feeding
The same types of feeders and waterers are used for both layers and broilers.
During their early life layer birds are fed as follows: *starter* - day-old to 6 weeks of age, *grower* – 6 to 8 weeks of age and *developer* – 8 to 15 weeks of age.

You will observe that layer birds start to lay eggs when they are about 4 to 5 months old. Therefore you should start feeding the layers *egg ration* or *laying ration* when they are 15 weeks of age. Continue to give them this feed throughout their life.
In addition, laying birds must be fed grit and ground oyster shells. Can you say why?
The grit helps to grind the food properly in the gizzard.
The oyster shells supply calcium which is necessary for good quality eggshells.

You may have observed that the egg yolk in farm eggs is pale yellow compared to the bright orange-yellow in common fowl eggs. Can you give an explanation for this?

It is because common fowls are generally allowed to run freely around our homes and eat green grass or herbage.

Substances in the green grass or herbage are responsible for the high vitamin A content which is found in the bright orange-yellow egg yolks of common fowls.

Therefore, it is advisable for us to hang small bundles of green grass or herbage for the layers to eat every day.

(iii) Pest and disease control

Pest and disease control measures are similar to those outlined for broilers.

Since layers are kept for a longer period than broilers it will be necessary to
— place more litter material on the floor:
— make greater use of medicines:
— observe the birds for pests and diseases:
— remove or isolate sick birds for treatment:
— cull or destroy birds which are unthrifty or not producing.

(iv) Record keeping

In addition to the kinds of records mentioned for broilers, egg records must be kept.

Now turn to your Workbook and do **Activity 37**.

(c) Management of rabbits
(i) Housing

We have learnt that rabbits are reared in hutches.

Observe that hutches are placed in well-ventilated buildings where they are protected from the elements.

Now observe the features of a rabbit hutch:
— sizes: 75 cm long, 60 cm wide and 45 cm high (can be used for rearing one adult rabbit or 2 young rabbits), or 120 cm long, 60 cm wide and 45 cm high (can be used for rearing 2 adult rabbits or 4 young rabbits);
— well-ventilated (wire cages or hutches are best).

(ii) Feeding

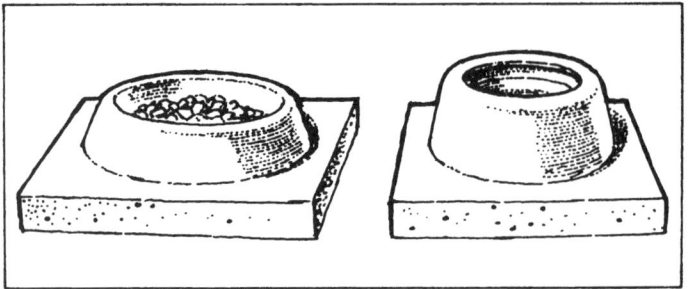

Observe that feeders and waterers are fixed in a concrete base. Why?
This prevents the containers from being turned over easily as the rabbits move about the hutch.

Can you name materials which are used for feeding rabbits?
Rabbits are fed concentrates, such as *rabbit ration* or *broiler starter*. They are also fed wilted herbage. Why?
Rabbits are normally fed clean, fresh feed twice daily, morning and afternoon, and they must be provided with fresh, clean water at all times.

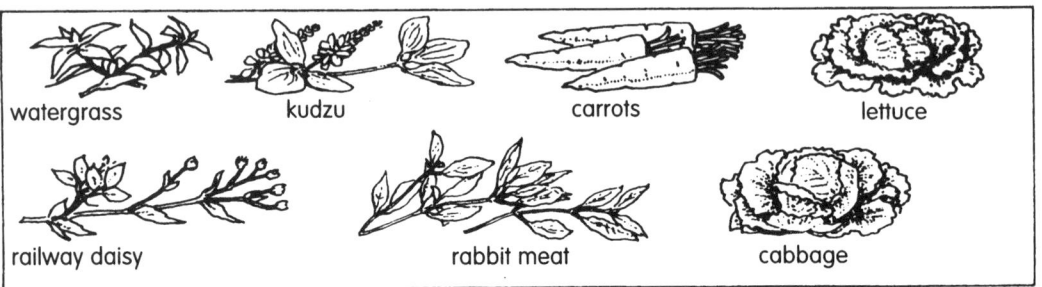

(iii) Pest and disease control
Proper sanitation is essential for pest and disease control.
Clean hutches, feeders and waterers daily and keep the surroundings clean.
Spread litter such as bagasse, straw and sawdust or wood shavings on the ground floor beneath the hutches. Why?
This helps to collect the droppings.
Stir the litter 2 or 3 times per week.
Remove and add fresh litter when necessary.

Examine rabbits for injuries, pests and diseases. Treat injured rabbits.
Rabbits also get *coccidiosis* especially if they are reared near poultry.
The treatment is the same as for poultry: use sulphur drugs (e.g. 5-Sulfas) in the drinking water.
You may observe that the nostrils of rabbits become somewhat clogged or runny with cold (mucus).
This causes them to sneeze a lot.
This disease in rabbits is known as *snuffles*. Eucalyptus oil is applied to their nostrils for control.

(iv) Record keeping
As for poultry, records are important in rabbit rearing, for example, breeds, dates of birth, dates of service, numbers of litters produced and so on.

Now do **Activity 38** in your Workbook.

(d) Management of goats
(i) Housing
Look carefully at a goat house and observe its features.

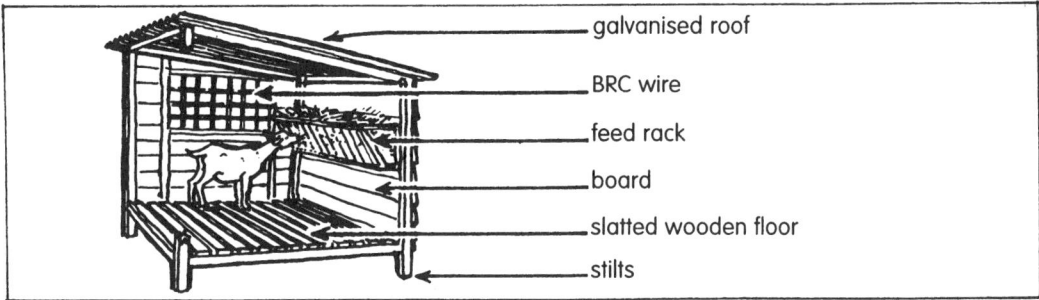

Notice that it gives protection to the goats from the elements and from thieves.
Notice also that the floor of the goat house is slatted and is raised from the ground. This allows the droppings to fall to the ground where they can easily be removed.
A goat house must also be well-ventilated.

(ii) Feeding
Observe the feed rack (for grass), feed tray (for concentrates) and bucket (for water).

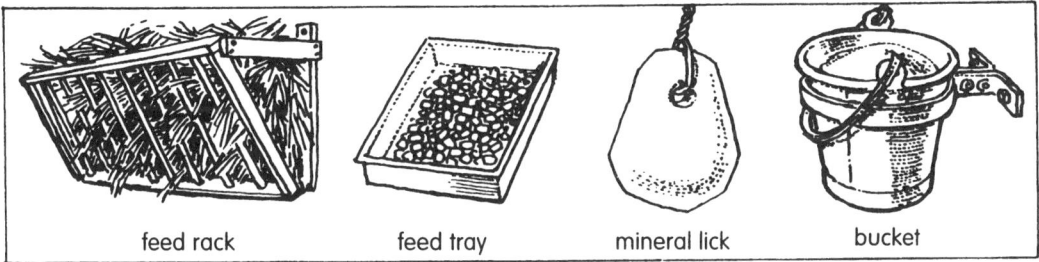

What materials are fed to goats?
Feed materials include:
— herbage or grass (goats are also allowed to graze);
— concentrates, e.g. *dairy ration*;
— mineral or salt lick which provides essential minerals.
Goats usually like dry areas, so allow them to graze or browse on well-drained pastures.

(iii) Pest and disease control
Proper sanitation is essential in goat rearing.
Sweep, scrub and wash the floor of the goat house daily.
Do the same to the ground floor. Remove all litter and droppings to a manure heap.
Wash the feed tray and bucket thoroughly each day.

Goats are fairly resistant to most pests and diseases. However, observe the animals daily for injuries and treat, if necessary. De-worm the animals using a worm medicine to get rid of intestinal parasites such as roundworms.

(iv) Record keeping
Keep records of the breeds, parents, dates of birth, dates of service, numbers of young produced and so on.

Now do **Activity 39** in your Workbook.

5 Bee-keeping

(a) The hive

Why do bee-keepers rear bees? In what are bees reared?
Bees are normally reared for honey in special boxes called hives.
They also provide us with wax.
The honey bee is known as a social insect.
Can you say why?
The honey bee is a social insect because it lives in a hive with thousands of other bees.

Have you ever looked inside a hive? What does it contain? The diagram below will help you.

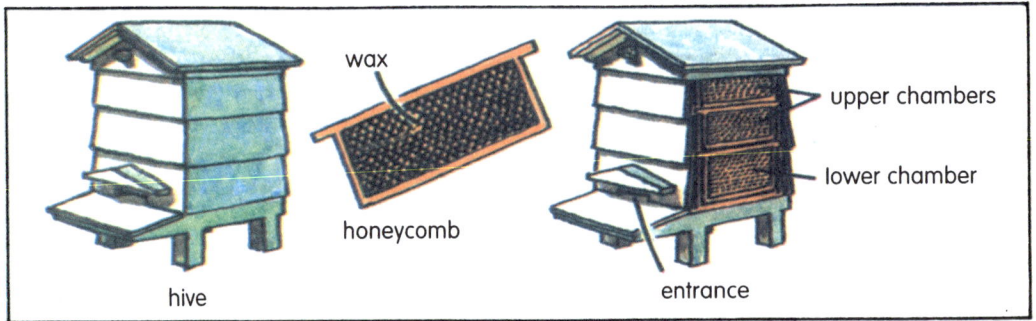

We must remember that bees will sting us if we disturb them.
Some types of bees are more fierce than others, for example, the Africanised bee.

It is important therefore for bee-keepers to wear protective clothing when attending to bee-hives.

Now let us look at the different kinds of bees in a hive and find out the work that they do.

(i) Queen bee
Every hive contains at least one queen bee. Her main function is to lay eggs, and she can sometimes lay about 1,500 eggs in one day. A queen bee will normally fight and kill other queen bees in the hive.

(ii) Drones
A hive contains hundreds of drones. Drones are male bees which do no work. As such they are said to be 'lazy'. The main function of drones is to mate with or fertilise the queen.

(iii) Workers
There are thousands of workers in a hive. The workers are sterile female bees which do all the work such as feeding the young, making wax, building cells, cleaning the hive, looking after the queen, guarding the hive, scouting and collecting pollen and nectar.

(b) The economic importance of bee-keeping
We have learnt that bees are important to us because they provide us with honey and wax.
Honey is not only important to us as a liquid food but is very useful in other industries, for example:
— the drug industry, for mixing in cough syrups; an alcoholic drink called *mead* can also be made from it;
— the food industry as a honey dip for fried chicken.
Wax is also used in many industries.

In addition honey bees are very valuable to us in agriculture because they pollinate the flowers of fruits and vegetables. As such, bees help to increase the yield of crops and in this way farmers can earn more profits.

Now turn to your Workbook and do **Activity 40**.

Unit 6 The business of agriculture

1 Agriculture as a business
(a) Factors and main objectives
Study the chart below. It shows that agriculture should be approached like any other business.

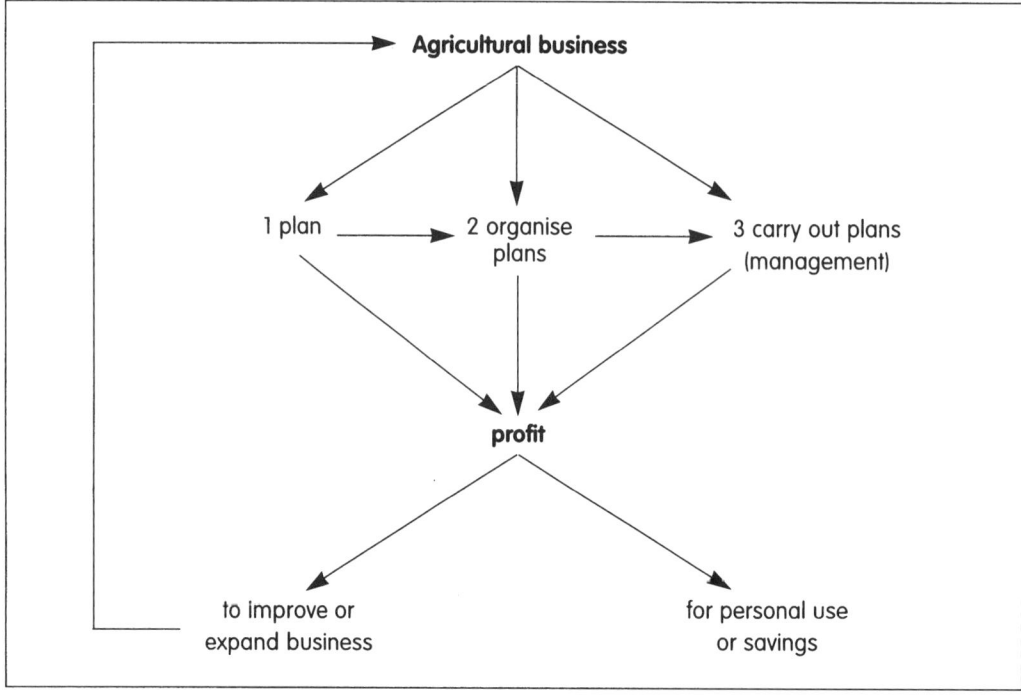

Observe that there are three factors which we must put into operation when running an agricultural business. Name them from the chart.
What is the main reason or objective of running an agri-business?
The chart shows that the main objective or motive for operating an agri-business is *to make profit* (money) and that we must follow the three factors if we wish to make *profit*.

What use is made of the profit?
Observe that some of the profit goes towards personal use and savings.
Some of the profit goes back into the business and is used primarily to improve or expand the business.

We can see therefore that agriculture is not simply a way of life. In fact, it is a *business* which requires careful planning, organisation and management.

(b) Importance of planning
Study the chart shown on the next page:

86

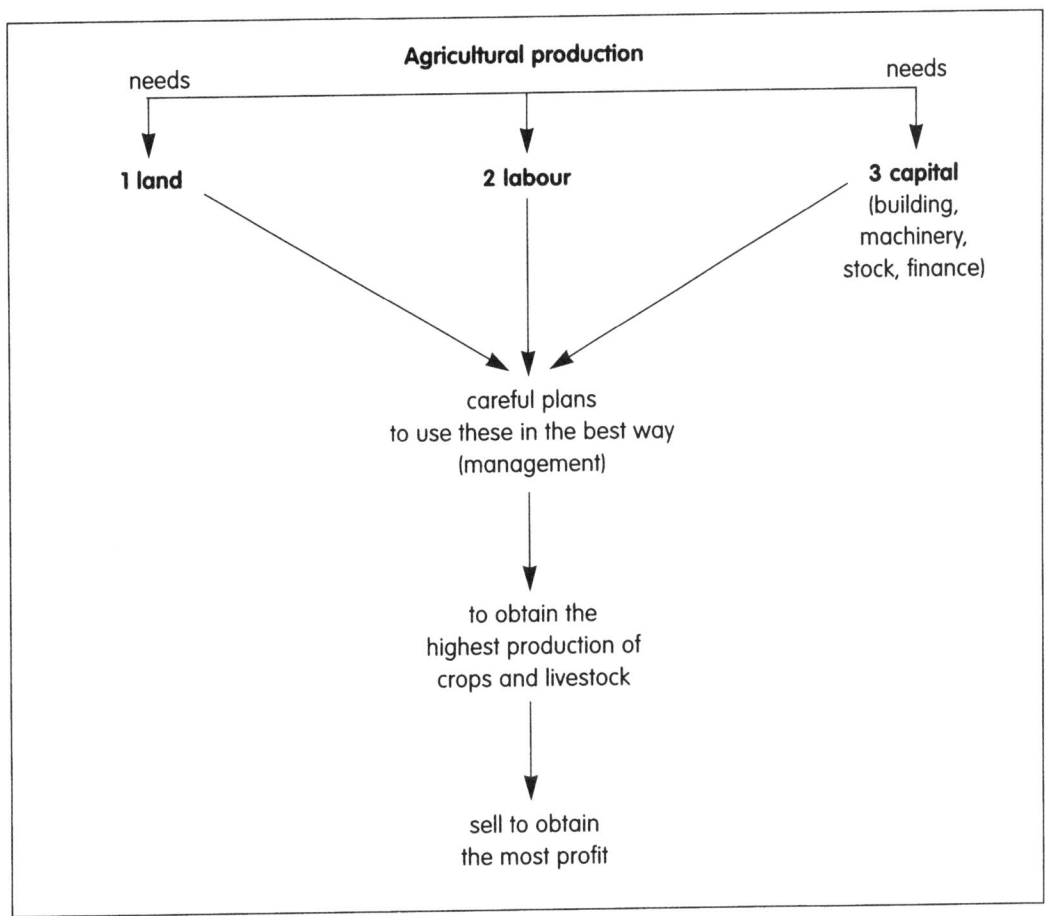

The chart shows that in order to produce crops or livestock a farmer needs land, labour and capital. These are known as factors of production or resources.

The farmer must be able to use these three resources in such a way that he does not just make a profit but makes the *most* profit.

How can the farmer make the *most* profit from his agricultural business?

With careful and proper planning a farmer can make the most profit from his agricultural business. For example, the farmer will have to *plan* and decide:

(i) What to produce?
— is it going to be tomatoes or cabbage? broilers or goats?

(ii) Why produce it?
— it may be because tomatoes will fetch a higher price than cabbages, or because broilers will mature faster than goats.

(iii) What variety or breed to produce?
— the farmer will, of course, plan on choosing a high-yielding variety of tomatoes which is suitable for the particular season, or an early maturing breed of broilers.

(iv) How much to produce?
— the farmer may decide perhaps to grow one hectare of tomatoes or to rear a batch of 2,000 broilers.

(v) When to produce?
— the farmer may start two or three months before Christmas in order to get the produce in time for Christmas when there is normally a great demand for chicken and tomatoes.

(vi) How to produce it?
— the farmer may use more machinery than labourers or may ask his family and children to help.

Proper planning, therefore, is the key to profitable agriculture. However, having made plans the farmer must organise and put the plans into action if he hopes to make the most profit from his agricultural business.

Now turn to your Workbook and do **Activity 41**.

(c) Simple financial records
You have already learnt that record keeping is important in agriculture. Can you name some kinds of farm records that should be kept? Check the chart below to find out.

Farm records
1 crop records
2 livestock records
3 financial records
4 personnel record (information, data on farm employees)
5 inventory (record of stock, equipment, assets)
6 farm diary (record of farm activities, projects – daily, weekly...)

In previous lessons you have learnt about crop records and livestock records. Check Units 4 and 5 to find out some of the information that should be recorded for crops and livestock.
In this lesson you will learn about financial records. Study the text below:

Financial records
(i) records of income or receipts
(ii) record of expenses or expenditure

Can you explain what financial records are?
As the name suggests, financial records are records of money or finance which is obtained and used by the farmer.

Notice that financial records include:

(i) Records of income: that is, money which is *received* by the farmer: includes loans and sales of all crops and livestock products.

(ii) Records of expenditure: that is, money which is *spent* by the farmer: includes money spent on land, housing, machinery, stock, feed, medication, pesticides, electricity or power, transportation and labour.

Why are financial records important in agriculture? Here are some reasons:
1 Financial records enable the farmer to know exactly how much *profit* or *loss* he is making from his farming business.
2 Financial records prevent the farmer from relying on his memory or from guessing; it is better to put records in black and white, that is, to write them down in a book.
3 Financial records help the farmer to keep regular *checks* on his income and expenses.
4 Financial records can enable the farmer to obtain a loan from the Agricultural Bank.
5 Financial records supply written proof and as such can be used to settle disputes or arguments.

Now, here is a simple way of keeping financial records. Carefully study the example below:

Date	Income (sales & receipts or loans)	$	¢	Date	Expenditure (purchases & expenses)	$	¢
3.1.80	Loan from Bank	5,000.00		7.1.80	1,000 day-old broiler chicks	420.00	
6.3.80	Sale of 2,000 kg chicken @ $2.50 kg	5,000.00		7.1.80	Vaccine & medicines	42.00	
				7.1.80	50 bags broiler starter	1,500.00	
				31.1.80	25 bags broiler starter	750.00	
				14.2.80	30 bags broiler finisher	900.00	
				10.3.80	Wages to farm attendant	1,000.00	
					Interest on loan	94.00	
	Total income	$10,000.00			Total expenditure	$4,706.00	

Examine the total income and the total expenditure.
Has the farmer made a profit or a loss?
How can you tell from the figures that he has made a profit?
It is because his income is greater than his expenditure.
Find out how much profit he has made, by subtracting his total expenditure from his total income.

Now do **Activity 42** in your Workbook.

2 Marketing of produce
Study the chart below to find out what marketing is:

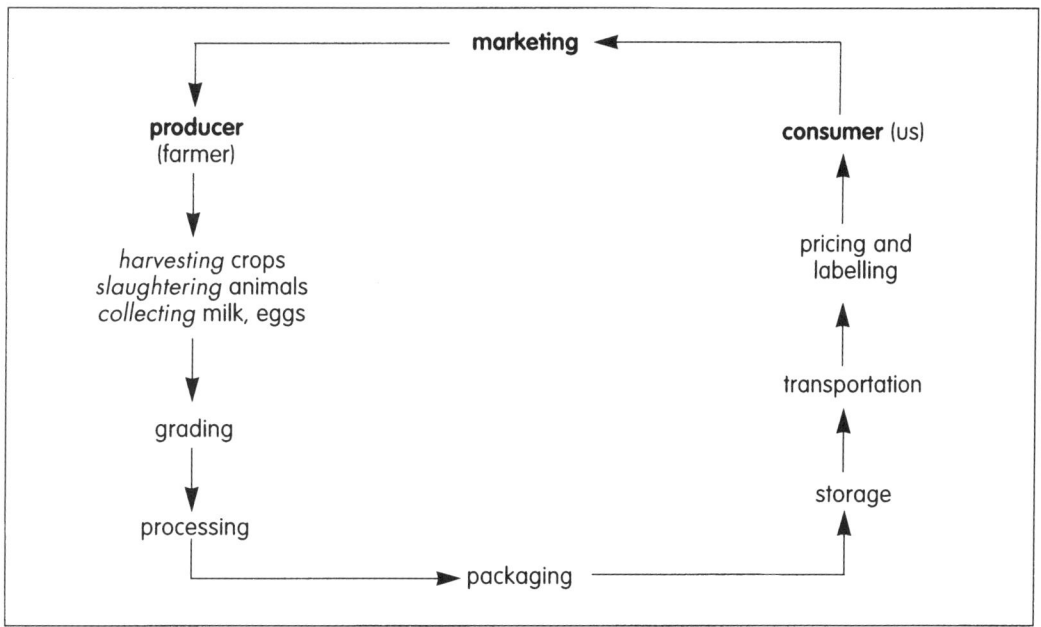

Observe that *marketing* includes all the *factors or operations* which are related to the flow of goods and services from the producer (farmer) to the consumer (us).
Check the chart and name some of the important factors which are involved in marketing.

Why are these factors necessary in marketing?

It is necessary to carry out operations or factors such as harvesting, grading, packaging and so on in order:

— to get produce to the consumer in the freshest possible state;
— to present goods in an attractive manner so that consumers will want to buy;
— to maintain desirable standards.

Now look closely at each marketing operation. Can you state the importance of each? Check the table below to find out.

Factor or operation	Importance
Harvesting	To prevent produce from spoilage in the field. If harvesting is not done in time, some produce will lose its attractive colour and nutritive value, and consumers will not want to buy it.
Collection of milk and eggs; slaughtering of animals	Proper methods must be used to maintain hygienic standards and to prevent spoilage. Consumers will not buy spoilt milk, meat or eggs, which means the farmer loses profits.
Grading	To separate the higher quality products from lower quality products. In this way the farmer may earn more profit because the better the quality, the higher the price of the product.
Processing	The product can keep for a longer period without spoilage; to enable consumers to obtain products out of season, such as canned pigeon peas.
Packaging	To make the product attractive and convenient for the consumer to handle.
Storage	To prevent spoilage; cold storage enables some products to remain fresh.
Transportation	To get the products quickly to the consumer in a fresh state.
Pricing and labelling	To enable consumers to recognise products and their prices at a glance. It saves the market vendor from having to answer questions about items and prices repeatedly.

Now turn to your Workbook and do **Activity 43**.

3 Industry and agriculture

(a) How industry helps agriculture

Look carefully at the pictures below:

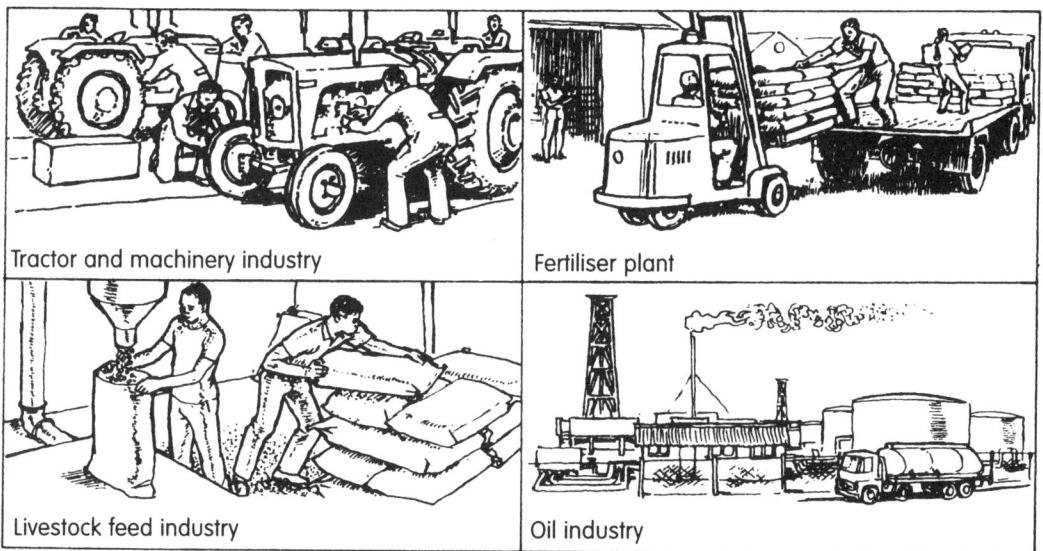

Do you recognise the different industries? Name them.

What, would you say, is an industry?
Observe that an industry is involved in manufacturing a particular product, for example, the fertiliser industry manufactures fertilisers.
You can see that each of the industries in the pictures is helpful to agriculture. Can you state in what way each industry is helpful to agriculture? The table below will help you.

Industry	Way in which it is helpful to agriculture
Tractor and machinery industry	provides machines for agriculture
Fertiliser industry	supplies fertiliser for crops
Livestock feed industry	provides feeds for farm animals
Oil industry	supplies money for studying agriculture, that is, for scholarship and research
Tourism industry	creates a demand for local foods

Now, can you name other industries and state how each one is helpful to agriculture?

(b) How agriculture helps industry
Look carefully at the two industries in the pictures below:

Can you state in what way each of these industries is dependent on agriculture?
You will observe that a sugar cane factory depends on sugar cane from agriculture for the manufacture of sugar, rum, molasses and bagasse.
Similarly, a meat processing factory depends on raw meat from farm animals such as cattle, pigs, poultry and so on.

Now, here are some other ways in which agriculture helps industry:
(i) Agriculture enables certain industries to *function*, such as manufacturing, processing, canning and bottling industries.
(ii) Agriculture provides raw materials to industries:
   — for *manufacturing* other products, such as cocoa for making chocolates, cotton for making clothing;
   — for *processing*, pork is processed into bacon, ham and sausage;
   — for *canning* and *bottling*, citrus, guava, pineapple, pigeon peas, peppers.
(iii) Agriculture provides food for the workers in industry.
(iv) Agriculture provides fresh foods and vegetables for the tourism industry.

Now find out the names of industries in your country which depend on agriculture.
In what way is agriculture helpful to those industries?

Now do **Activity 44** in your Workbook.

Unit 7 Agencies which help farmers

Government and non-government agencies

Every day our farmers have more mouths to feed because of the rapidly increasing population in our country.

With our increasing population, more and more of our agricultural land is being used up for housing, roads, schools and recreational facilities.

Because of constant cultivation and in some cases poor management, our land is becoming less fertile. Therefore, the task of providing more food for our nation is a difficult one for our farmers. Therefore our farmers need lots of assistance.

Can you name some of the agencies which help our farmers? Find out from the chart below:

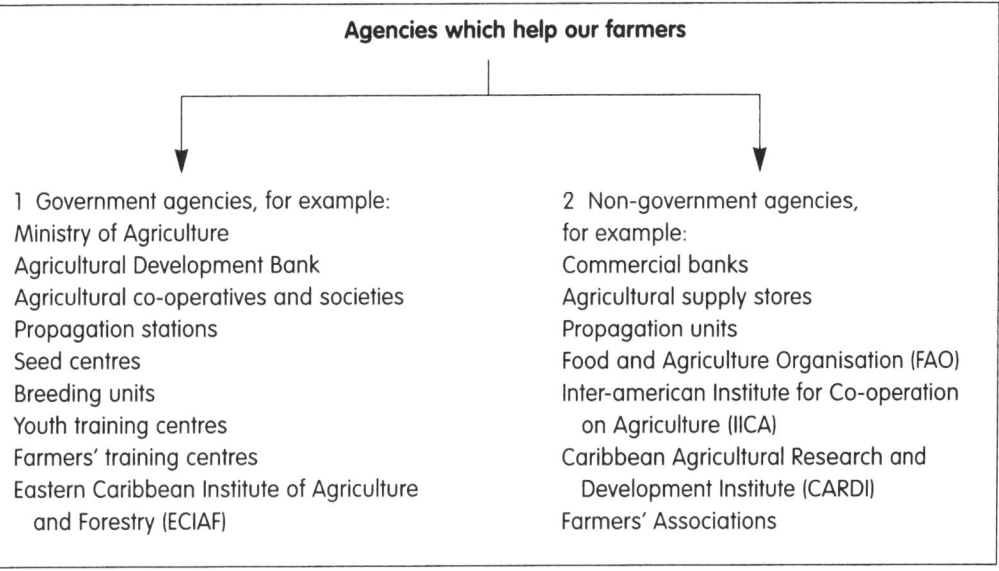

Observe that there are two main groups of agencies which help our farmers – (i) government agencies and (ii) non-government agencies.

Name some examples from each group from the chart.
With the aid of your teacher name agencies in your country which belong to each group.
Now can you explain how these two groups of agencies help our farmers? Check below:

Kind of help given to farmers by agencies

1 Subsidies (money or financial assistance)
2 Loans at low interest rates
3 New varieties of planting materials and improved breeds of livestock
4 Services such as soil testing; A.I. (artificial insemination of cows); veterinary (vet.) service
5 Technical advice from extension officers
6 Training courses in farming

Let us carefully study the table below which gives us the names of some important agencies and tells us how each agency is helpful to our farmers.

Name of agency	Government or non-government	How it is helpful to farmers
1 Ministry of Agriculture	Government	gives *subsidies* to farmers for example: — for land preparation — for growing vegetables and other food crops — for growing plantation crops such as rice, bananas, cocoa, coffee — for purchasing farm equipment. — for purchasing fertilisers
2 Farmers' training centres	Government	provide *training courses* for farmers, in, for example: — dairy farming — pig farming — vegetable growing, and so on
3 Agricultural Extension Department	Government	provides technical *advice* to farmers through agricultural extension officers, on — improved methods of growing crops — new methods of rearing livestock — how to prune coffee, cocoa — how to rear calves, piglets
4 Experimental and research stations	Government	— do soil testing for farmers — help to solve farmers' problems through research
5 Breeding units	Government	— provide a breeding service for farmers' livestock — provide veterinary (vet.) service
6 Marketing agency	Government	— provides collection depots for farmers' products — provides marketing depots which sell agricultural supplies to farmers
7 Agricultural Development Bank	Government	— provides *loans* to farmers at low interest rates
8 Agricultural co-operatives and societies	Government	— provide *loans* to individual farmers at low interest rates — assist groups of farmers to operate co-operative agricultural shops
9 Youth training centres	Government	— train young men and women to practise agriculture
10 Propagation stations	Government	— provide planting materials for farmers such as cocoa plants, citrus plants, plants of other fruits
11 Seed centres e.g. Chaguaramas, Trinidad	Government	— provide seed materials for farmers, such as corn, pigeon peas, etc.

Name of agency	Government or non-government	How it is helpful to farmers
12 Livestock stations e.g. Aripo, Trinidad	Government	— provide young and growing livestock for farmers
13 University of the West Indies (UWI), St Augustine, Trinidad	Government (Regional)	— provides training for students in agriculture — does research to find solutions to farmers' problems — supplies new agricultural methods to farmers
14 Caribbean Agricultural Research and Development Institute (CARDI)	Government (Regional)	— does research in crops and livestock — finds solutions to farmers' problems — supplies new agricultural methods to farmers
15 Eastern Caribbean Institute of Agriculture and Forestry, Centeno, Trinidad	Government	— provides training for young men and women in agriculture
Sir Arthur Lewis Community College, St Lucia	Government	— as above
Guyana School of Agriculture	Government	— as above
16 Commercial banks	Non-government	— provide *loans* at higher interest rates to farmers
17 Agricultural supply stores	Non-government	— sell agricultural supplies to farmers — provide *advice* to farmers
18 Inter-american Institute for Co-operation on Agriculture (IICA)	Non-government	— does agricultural research — finds solutions to farming problems — provides technical assistance to farmers
19 Food and Agriculture Organisation (FAO)	Non-government	— finds solutions to farming problems — provides technical and financial assistance for agricultural development projects — funds training programmes
20 Farmers' Associations	Non-government	— provides support and bargaining power to farmers

Now, with the aid and guidance of your teacher, name agricultural agencies in your country and state how each one is helpful to farmers.

Turn to your Workbook and do **Activity 45**.